HISTORY OF BROADCASTING: RADIO TO TELEVISION

HISTORY OF BROADCASTING: Radio to Television

ADVISORY EDITOR

Dr. Christopher Sterling, Temple University

EDITORIAL BOARD

Dr. Marvin R. Bensman, Memphis State University
Dr. Joseph Berman, University of Kentucky
Dr. John M. Kittross, Temple University

First Principles of Television

ALFRED DINSDALE

ARNO PRESS and THE NEW YORK TIMES
New York • 1971

Reprint Edition 1971 by Arno Press Inc.

LC# 76-161141
ISBN 0-405-03562-4

HISTORY OF BROADCASTING: RADIO TO TELEVISION
ISBN for complete set: 0-405-03555-1
See last pages of this volume for titles.

Manufactured in the United States of America

FIRST PRINCIPLES OF TELEVISION

FIRST PRINCIPLES OF TELEVISION

BY

A. DINSDALE, M.I.R.E.

(FORMERLY EDITOR OF THE *TELEVISION MAGAZINE*)
MEMBER OF COUNCIL AND EXECUTIVE COMMITTEE OF THE TELEVISION SOCIETY
AUTHOR OF *TELEVISION*

LONDON
CHAPMAN & HALL, LTD.
11 HENRIETTA ST., W.C. 2
1932

PRINTED IN GREAT BRITAIN BY THE ABERDEEN UNIVERSITY PRESS, ABERDEEN, SCOTLAND

TO
J. E. D.

PREFACE.

THIS book does not describe *every* system of television that has ever been experimented with; the list is so long that the reader would only become bored. What I have endeavoured to do is to pick out, in the earlier chapters of the book, the most outstanding proposals of the early workers, in order to provide an adequate background for the subject. In the later chapters, dealing with contemporaneous work, I have described the apparatus and methods of those workers who, at the time of writing, have gone farthest along their own chosen line. Little or no attention has been given to those workers, who are more or less duplicating the methods and results of others as a preliminary study to more advanced and original work. By confining myself to the work of the most distinguished experimenters (judged by results and methods), I feel that I have justified the title of this book and achieved my objective, which is to expound all the *principles* of television (as at present known), and describe the apparatus with which present results are being achieved.

Progress in television is now so rapid that it is impossible for any book on the subject to be absolutely up to date, but it is my hope and belief that this book will serve as a work of reference which will enable the student readily to understand any future system of television which may make use of combinations or variants of the principles herein described. There is, of course, always the possibility that some entirely new principle may be discovered and successfully applied at any time.

As an example of a noteworthy variant, only a few days ago, I was invited to a demonstration in New York by U. A. Sanabria, a young Chicago experimenter, who projected close-up images on to a screen six feet square which were better than anything I have ever seen. This must be taken as superseding comparative statements made later in this book. Sanabria uses at both transmitter and receiver a triple spiral disc, each spiral containing fifteen holes only, the holes in successive spirals being offset in such a manner that the light-spots overlap one another on the screen. The total number of holes is therefore only forty-five, and the speed of transmission is fifteen pictures per second.

At the receiver Sanabria uses a special neon arc which he has developed, and which provides a degree of illumination on the receiving screen which is about half that of a moving picture screen. Because of the unusual arrangement of the holes in the discs, and by reason also of a special transmitter amplifier which steepens the wave-front of the signal impulses, there is an almost complete absence of flicker, and the detail of the image is quite as good as that provided by the average home cinema. Scanning lines are much less apparent than in any other system which I have been privileged to examine.

Critics may assail me for thus including in my preface what might be termed a technical afterthought, but I want this work to be as up to date as possible, and this digression lends point to my second paragraph.

Throughout this book I have endeavoured to avoid becoming too technical, so as to make the work available to as wide a circle of readers as possible. Anyone possessing some knowledge of wireless and electricity should experience no difficulty in following the text, which is freely illustrated with both diagrams and photographs.

In the preparation of a book of this type, many sources of information have to be drawn upon, and

PREFACE

I desire to express my indebtedness to many of the workers mentioned herein for their courtesy and co-operation in showing me their apparatus and results, and for supplying me liberally with information. Special acknowledgment must be made to the Television Press, Ltd., for permission to use valuable matter which has already appeared in *Television*. I desire to acknowledge, also, the following sources of information and/or diagrams: Bell System publications, Figs. 55-63, 92, 108-111; P. T. Farnsworth, Figs. 119-130; G.E.C. (Great Britain), Figs. 20 and 21; G.E.C. (U.S.A.), Figs. 112 and 113; *Projection Engineering*, Figs. 114-118; Radiovisor Parent, Ltd., Fig. 18; *Television*, Figs. 16, 17, 19, 23, 24, 47, 50, 77, 93, 94, 96, 97, 99-101, 105, 106; *Wireless World*, Fig. 104. Where the origin is known, the photographic illustrations are individually credited.

With these few introductory paragraphs, I commend what follows to the attention of my readers, and trust that it will meet with a favourable reception.

A. D.

NEW YORK, 1932.

CONTENTS.

CHAPTER I.

INTRODUCTION 1
Definition. Importance of Sight in Communication. Differentiation between Television and Phototelegraphy. Principal Requirements. Simple Analogy.

CHAPTER II.

ELEMENTARY CONSIDERATIONS 7
Simple Optical Systems. The Human Eye. Properties of Selenium. The Experiments of Rignoux, Fournier, and Ruhmer. Exploring an Image.

CHAPTER III.

LIGHT-SENSITIVE DEVICES 18
Simple Selenium Cell. Nature of Selenium. Time Lag. Radiovisor Bridge. Photoelectric Effect. Vacuum and Gas-filled Photo Cells. Amplification. Zworykin's Combination Cell. Shadowgraphs and Television Differentiated. Requirements for Television.

CHAPTER IV.

SOME EARLY TELEVISION EXPERIMENTS 33
Szczepanik's Vibrating Mirror System. Rosing's Mirror Drums and Braun Tube. Cathode Rays. Mihaly's Oscillographs. Belin and Holweck's Cathode Ray Oscillograph.

CHAPTER V.

THE JENKINS SYSTEM 46
Jenkins Prismatic Discs. Glow Lamps. The Moore Tube. Animated Shadowgraph Transmitter. Lens Disc. " Radiomovie " Receiver. Jenkins Drum Scanner. 4-Electrode Neon Tube. Advantages of Drum Scanning.

CHAPTER VI.

THE BAIRD SYSTEM 58
Early Efforts. Nipkow Disc Scanner. Spot-Light Transmitter. Flat Plate Neons. Lens Disc Scanner. Flood-Light Transmitter. Optical Lever. Relative Motion. Multiple Channels.

CONTENTS

CHAPTER VII.

THE BAIRD SYSTEM—*continued* 71
 Phonovision. Listening to a Face. Phonovisor v. Talkies. Noctovision. The Spectrum. Generating Infra-red Rays. Penetrative Qualities of Infra-red Rays. Noctovisor as Aid to Navigation. Television by Daylight, in Colours, and in Stereoscopic Relief. Enlarging the Received Image.

CHAPTER VIII.

THE BELL SYSTEM 87
 1927 Monochrome Demonstration. Circuits and Methods Employed. Cells and their Arrangement. Details of Large Screen. Method of Illuminating and Modulating Large Grid Neon. Daylight Demonstration. Direct Scanning. Optical Arrangements. Colour Television. Sodium Cell. Arrangement of Cell Cabinet. Transmitter Amplifier Circuits. Arrangement of Glow Lamps, Filters, and Mirrors at Receiver. Receiver Amplifiers. Adjustment Problems.

CHAPTER IX.

METHODS OF SYNCHRONISM 107
 Independent Control. Synchronising in Phototelegraphy. Electrically-Driven Tuning-Forks. Crystal-Controlled Valve Oscillators. Direct Control. Synchronous Motors and Phonic Wheels. Defects Caused by Lack of Synchronism. Isochronism and Synchronism Defined. Hunting Propensities of Motors. Applying Synchronous Motors to Check Hunting. Method of Phasing the Image. Transmission, Amplification, and Application of Isochronising Impulses. Using A.C. Mains.

CHAPTER X.

IMAGE STRUCTURE 121
 Methods of Signalling. Structure of Half-Tone Reproductions. The Dot Theory of Television. "Picture Points." Aperture Distortion. Picture Ratio. Horizontal v. Vertical Scanning.

CHAPTER XI.

TRANSMISSION CHANNELS 135
 Frequency Limitations of Existing Channels. Use of Submarine Cables and Long Wireless Waves Impossible at Present. Distances Covered to Date by Television. Difficulties on Short Waves. Medium Waves Best. The Sideband Theory. All Wavebands Overcrowded. The Stenode Radiostat Explained. Bell System Experiments with Multiplex Transmission, and Conclusions. Stenode Radiostat Principle Promises Vast Extension of Transmission Channels and Band Widths. Wire Channels Best at Present.

CHAPTER XII.

THE PRESENT STATE OF THE ART IN GERMANY . . . 157
 Fernseh, A.G., Arrangements. German Post Office Apparatus. Aims of Post Office. Karolus-Telefunken System. Telefunken Short-Wave Experiments. The Mihaly System. Von Ardenne's Experiments. Review of Position in Germany.

CONTENTS

CHAPTER XIII.

THE PRESENT STATE OF THE ART IN ENGLAND . . . 170

Television Broadcasting Facilities. Baird Studio and Control Room. Baird "Televisor" Receiver. Cog Wheel Automatic Synchroniser. Relay Synchroniser. Baird's Large Screen. Tele-Talkies. Correcting Faulty Reception. Echo Images. Review of Position in England.

CHAPTER XIV.

THE PRESENT STATE OF THE ART IN AMERICA. MECHANICAL SYSTEMS 189

Jenkins Radiovisors and Broadcasts. Bell System Two-Way Television. Method of Synchronising. The Acoustic System. Alexanderson's Large Screen. Appendix — Wavelength and Power Assignments for Experimental Television Broadcasts in America.

CHAPTER XV.

THE PRESENT STATE OF THE ART IN AMERICA (*continued*). CATHODE RAY SYSTEMS 206

Zworykin's Film Transmitter. Kinescope Receiver. Method of Synchronising Pros and Cons of Cathode Ray Method. Farnsworth's Dissector Tube. Electron Image. Magnetic Focusing. Construction and Operation of Dissector Tube. Receiving Oscillite. Method of Synchronising. Comparisons. Review of Position in America.

CHAPTER XVI.

CONCLUSIONS 229

Various Opinions Concerning Television. Possibilities of Television as an Entertainment Medium, and Requirements. Can it be Done? Unlimited Scope for the Amateur. Does the Public Want Home Television? Television as a Stepping-Stone, and as an Adjunct to the Motion Picture Industry. Other Applications. Noctovision and Television as Aids to Navigation. The International Newspaper.

INDEX 239

LIST OF PLATES.

	FACING PAGE
Baird's Original Transmitter	6
The Radiovisor Bridge	12
Zworykin's Combination Photocell and Valve	24
Belin's Apparatus	30
Jenkins Lens and Prismatic Discs	36
Flat Plate Neon	44
Early Baird Apparatus	48
Baird's Noctovision Transmitter	54
Baird's Improved "Noctovisor"	60
Baird's Daylight Television Transmitter	66
Transatlantic Transmission	72
Bell System Transmitter and Receiver	78
Bell System Large Screen	84
Alexanderson's Seven Spot Projector	92
American Transmitter and Receiver	96
Electrically Driven Tuning-Fork	102
Simple Synchronous Motor	108
Karolus-Telefunken Transmitter	120
Fig. 76.—Three half-tone blocks	124
Karolus-Telefunken Transmitter. Another Form	126
Karolus-Telefunken Large Screen Receiver	132
Telefunken Combination Receiver	138
Mihaly Transmitter and Receiver	144
Telehor Co.'s 1930 Receiver	150
Baird Co.'s First Studio	156
Control Room, Baird Co.	162
Baird "Televisor" Receiver	168
Baird Synchronising Gear	174
Baird's Large Screen	180
Simple Jenkins Radiovisor	186
Jenkins Radiovisor, Console Model	194
Bell System Two-Way Television Booth	198
Bell Two-Way Equipment	204
Water-Cooled Neon Tube	210
G.E.C. Transmitter	216
Alexanderson's Large Screen Apparatus	222
Farnsworth's Cathode Ray Tubes	228
Farnsworth Experimental Receiver	234

CHAPTER I.

INTRODUCTION.

Definition. Importance of Sight in Communication. Differentiation between Television and Phototelegraphy. Principal Requirements. Simple Analogy.

THE last thirty years have seen the growth and popularisation of a host of scientific wonders, outstanding examples of which are the cinematograph, the gramophone, the telephone, wireless telegraphy and telephony, broadcasting, the aeroplane and the airship, and the automobile.

No one but a Jules Verne could have visualised, a hundred years ago, that to-day we would be able, by means of a simple instrument to be found in every business office and a very large number of homes, to converse freely with friends and business associates in almost every part of the civilised world. Yet, to-day, this can be done with such ease that we accept the fact as part of our everyday lives, and think no more about it.

We take but little interest in the technical details of most of our wonderful inventions, unless, as in the cases of motoring and broadcast reception, the mechanism is under our own direct control. We have nothing to do with the intricate mechanism of the telephone service, so we take no interest in it.

In the case of the latest scientific marvel, television, its promise to add the gift of sight to our existing broadcasting facilities has fired the public imagination. As soon as it was announced by Baird, five years ago, that he had solved the problem of television there was an immediate interest taken by the public, and an increasing demand arose for technical articles and books on the subject, a demand which the present author, who was first in the field, has done much to meet.

The word "television" is derived from the Greek word *tele*, which means "at a distance," and the Latin verb *videre*, "to see." Television literally means, therefore, "seeing at a distance." We can, of course, see at a distance by means of a telescope, or a pair of field glasses, but the range of such optical aids is, as everybody knows, limited by the curvature of the earth and by atmospheric conditions. Television will give us a kind of electric telescope of unlimited range, and instantaneous in action. In other words, by means of television we shall be able to watch events taking place one hundred or one thousand miles away, and see them instantly, as they occur, just as now, by means of broadcasting, we can instantly *hear* what is taking place in a distant studio.

In the interests of a pure language, philologists may not like the word " television," because it is derived from two different languages. But it has been coined, and is now in everyday use, so, we venture to think, it will continue to be used.

Television, in the stage of development which it has reached to date, may be described as a combination of practices known to several sciences and arts. Mechanics, optics, chemistry, electrical and wireless engineering have all contributed their quota of knowledge which, properly applied, has resulted in the achievement of television. It is but one more instance of the fact, rapidly being recognised, that the various branches of science are not so widely separated and independent of one another as it was once customary to believe them to be.

When Alexander Graham Bell first discovered the telephone, his contemporary fellow-scientists dubbed it an " interesting toy," and let it go at that. No one but the famous inventor himself had the vision to foresee the usefulness of his discovery and its enormous potentialities for service to mankind.

Such is the general lot of pioneers. Too often their work is not appreciated at its full value until they have long since departed from this world. In the case of the more fortunate, even they have to contend with public scepticism and apathy, and perhaps opposition, before the general public finally acquires a sufficient knowledge of the latest invention and its potentialities to accept and make use of it. Unless this happens within the lifetime of the inventor, there is no profit in the business of inventing—at least, not for him.

Let us consider for a moment, therefore, the significance and potential usefulness of television. Granted that we can see what is happening at a distance, instantaneously, how can we make practical use of this ability ?

One of the most vital factors in the progress of the human

INTRODUCTION

race, from the beginning of time, has been communication. It is of the utmost importance to us that we should be able to establish communication with our fellow-men, and one of the earliest methods of doing so, and still the most effective, is by word of mouth.

Direct communication by word of mouth, to be thoroughly effective, involves something more than the mere employment of our vocal chords, and a language. It is not generally realised what a great part is played in oral communication by facial expression. Shut your eyes, or talk to a man in pitch darkness, and, unless you know him very well indeed, the intelligence conveyed by facial expression will soon be missed.

Consider the telephone. Apart from the physical imperfections of the instrument, it is surprising how often failure fully to grasp the distant speaker's import is due to our inability to *see* him. Thus, when we endeavour to receive a communication by telephone, we have to rely upon but one of our senses, that of hearing.

Similarly, take the " gags " of certain comedians. The actual words they use, when set down in cold type, or broadcast, frequently appear to be completely lacking in humour and pointless. But when spoken across the footlights, simultaneously with suitable facial expressions, they produce roars of laughter amongst the audience. In the case of broadcast humour, the studio audience, which can see the speaker, can often be heard to express its merriment at something which, heard only through the medium of a loud speaker, does not appear to be in the least amusing.

In the fields of radio broadcasting and gramophone entertainment we revert to conditions similar to those under which we labour when conducting a telephone conversation. We are being supplied with what may be termed blindfold entertainment. We hear what is going on in the broadcasting studio, or what has been impressed on the record, as the case may be, but the visualisation of the scene in the studio is left to our imagination. There are some who prefer it so, but the instant and phenomenal success of the " talkies " undoubtedly indicates that most people prefer reality to imagination.

From what has been written, it is evident that when it comes to communicating human intelligence, sight is as important a factor as hearing, if not more so, and in order to convey *personality*, sight is essential. It is the interaction of intonation, of voice and facial expression which conveys that elusive component of information, or intelligence, which we call personality, and it is a fact that very very few people indeed are so gifted that they can convey an adequate impression of their personality to a complete stranger by voice alone (e.g. via the microphone). Hence the imperative need for simultaneous sight.

There is surely an important application in the field of communication, therefore, for any device which will reinforce our receptive powers by enabling us to utilise the sense of sight as well as that of hearing, when we endeavour to communicate over great distances. That this is recognised by the general public is very evident from the great interest which it has taken, and continues to take in the development of this latest invention.

It is perhaps natural that the most obvious application for television should have been seized upon for development and exploitation first, and steps taken to ally it with oral broadcasting. However, broadcasting is but one phase, and a relatively unimportant one, of modern electrical communications. In some respects television may be described as the fastest means of communication known to man, and many of its processes and methods might very well be applied to other communication services for the rapid transmission and reception of visual or visible intelligence. These applications will be discussed in a later chapter.

Before proceeding to an examination of the technicalities of television, it is perhaps advisable to clear up a misunderstanding which still persists. Television is not the same thing as phototelegraphy. The former means the instantaneous sight of distant moving objects. The latter, as the name implies, means the transmission, by electrical means, of photographs, drawings, or facsimiles of documents, etc. At the receiving end the incoming information is directly recorded either on specially prepared paper, or, in the case of modern systems, on photographic film, from which copies may be made in the usual photographic manner.

Phototelegraphy is the older art of the two. By means of it, newspapers send topical photographs from one city, or continent, to another in time to print them immediately after the occurrence of the event which they depict, thus providing their readers with pictorial as well as typographical information with the minimum of delay. Business firms employ the world's

INTRODUCTION

phototelegraphy services to send in a shorter space of time than could be accomplished even by a specially chartered aeroplane, facsimile reproductions of important documents to branch offices or other firms with whom they may be doing business. In the case of important telegrams where a telegraphic error might be fatal, absolute accuracy can be secured by sending a facsimile reproduction of the original written or typed telegraph form.

These are some of the uses of the phototelegraphy services which are at present being rapidly developed. Their important characteristic is that they provide a record on paper of what was sent. Television does not attempt to provide a record. It provides a fleeting stimulus to the eye of the observer who is looking into his receiver; he is, as it were, conveyed to the distant transmitter and enabled to watch what is happening before it.

In the laboratory, by phototelegraphic methods, a picture measuring 5 inches by 4 inches can be transmitted from one place to another in the space of time of two or three seconds. Over much greater distances, outside the laboratory, using commercial apparatus, the time taken is very much longer, and may be as much as half an hour, depending on various conditions such as the distance to be covered, and the system of transmission being employed. In television, in order to give the observer the illusion of natural movement, it is necessary to transmit at least ten complete pictures, or successive views of the distant scene *per second*. It will be seen, therefore, that whereas the speed factor is relatively unimportant in the case of phototelegraphy, it is of vital importance in television, and, as will be seen later, it is the speed factor which has delayed the successful achievement of television for so long, and still presents many serious problems which will have to be overcome before further progress can be made.

The principal requirements for the achievement of television may be stated as follows:

At the transmitter, there must be provided (*a*) means for analysing the scene to be transmitted, or an image of it, into small areas, and (*b*) means for transforming the light values which represent these areas into electrical impulses of proportionate intensity; between transmitter and receiver a wire or wireless circuit, link, or channel must be provided through which the electrical impulses can be sent to the receiver; at the receiver there must be provided (*a*) means for retransforming the incoming electrical impulses back into light impulses of proportionate intensity, and (*b*) means for synthesising, out of the light impulses, an image which shall be as nearly as possible an exact replica of the scene or image at the transmitter. Means must also be provided to insure that the analysing and synthesising devices function exactly in synchronism, or in step with one another. The most important requirement of all these means is that they must function instantaneously, or so nearly instantaneously that the eye of the observer cannot detect in the received image any trace of lag or sluggishness in any component.

Where television is allied to sound broadcasting, and transmitted by wireless, a television transmitter *as a unit* may be compared to a microphone, which transforms the sound of a human voice or an orchestra into electrical impulses which are passed on to the wireless transmitter and broadcast, only instead of dealing with sound waves the television transmitter deals with light waves and transforms *them* into electrical impulses which are then broadcast in a precisely similar manner. A television receiver, considered also as a unit, may be compared to a loud speaker, which takes the electrical impulses from a broadcast receiver and retransforms them into sound impulses which are (or should be !) an exact reproduction of the speech or music being produced before the microphone, only instead of transforming the output of the wireless receiver into sound impulses the television receiver transforms it into light impulses, so arranged as to form an exact reproduction of the lights and shades of the scene before the distant transmitter.

Having thus introduced our subject, we will proceed to an examination of the various principles, means, and technical methods which have been employed, both unsuccessfully and successfully, to achieve television.

[*Courtesy: Baird Television Ltd.*

J. L. Baird's original transmitting apparatus, by means of which television by indirect light was first accomplished in 1925. The apparatus is now in the Science Museum, South Kensington, London.

[*To face page* 6.

CHAPTER II.

ELEMENTARY CONSIDERATIONS.

Simple Optical Systems. The Human Eye. Properties of Selenium. The Experiments of Rignoux, Fournier, and Ruhmer. Exploring an Image.

ONE cannot proceed very far in the study of television without coming in contact with optical systems of one sort or another, used either as image-forming or as image-casting devices. Mirrors, lenses, and prisms are all used in modern television apparatus, so it is perhaps as well, at the outset, to know something of the elementary characteristics of these things. A complete review of the science of optics is, of course, quite impossible here, and those who are interested to know something more about it are referred to the many excellent textbooks on the subject.

One of the first laws of optics deals with the transmission of light, and states that light travels in straight lines in any homogeneous medium. In passing from one medium to another, however (e.g. from air to glass), a ray of light is bent, or deviated, owing to the refraction which takes place at the surface which separates the two media. This is shown in Fig. 1, where a ray of light COD is shown passing from air through glass. The angle COA of the incident ray CO to the perpendicular or normal, AOB, is called the angle of incidence, and the angle DOB made by the refracted ray is called the angle of refraction. It is because of refraction that lenses are possible.

The next important law deals with the intensity of the light emanating from a point source, and states that the intensity decreases inversely as the square of the distance from the source. In other words, if the intensity is 4 at a distance of 1 foot, it will be reduced to 1 at a distance of 2 feet. This law does not hold good if the light source is larger than a mathematical point, or if the source is enclosed in a reflector; in the latter case the rate of decrease of intensity depends upon the reflector.

8   FIRST PRINCIPLES OF TELEVISION

As we shall see later on, reflected light is of great importance in television. All non-self-luminous bodies reflect light to a greater or less degree. One of the best reflecting bodies, and at the same time one of the simplest image-forming devices is a mirror, and for rays of light striking a mirror there is an optical law which states that the angle of incidence is equal to the angle of reflection, as shown in Fig. 2. This law does not hold good for any other than a mirror or other similarly highly polished surface.

For unpolished or matt reflecting surfaces, such as blotting paper or white fabric, the reflection follows Lambert's cosine

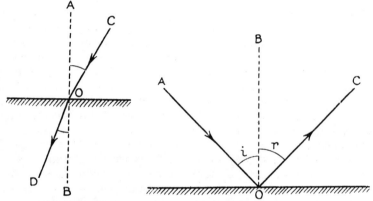

FIG. 1.—In passing from one medium to another a ray of light is bent. AOC is the angle of incidence and DOB is the angle of refraction.

FIG. 2.—In the case of a mirror the angle of incidence AOB is equal to the angle reflection BOC.

law, which can best be understood by reference to Fig. 3, where the length of the arrows terminating on the circumference of the circle represents the amount of light which is reflected at any given angle. In televising a scene by the flood-light system (which will be described in detail later on), the location of the lights and the choice of reflectors is of the greatest importance.

Reference has been made to the image-forming property of a plane mirror. In this connection the reader has no doubt noted that the image seen in a mirror appears to be as far *behind* the mirror as the object is in front of it. No rays of light can actually penetrate the mirror, so that, in fact, no

ELEMENTARY CONSIDERATIONS

rays exist where the image appears to be, and such an image is known as a *virtual* image.

In the case of curved mirrors, however, the laws which govern their use are somewhat more complicated. In Fig. 4 a ray of light, A, strikes a concave mirror and is reflected in accordance with the reflection law already quoted, because to a ray of light of pin-point area the surface of the mirror acts as if it were plane, or flat. Ray B will be similarly reflected, and the point at which the two reflected rays intersect is called the focal point of the mirror. This focal point will be found to be half-way between the mirror and its centre of curvature, C.

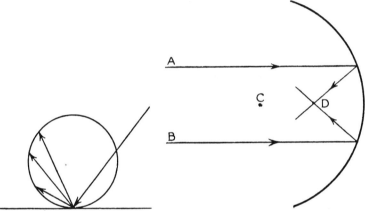

FIG. 3.—Illustrating Lambert's cosine law for diffusely reflecting surfaces.

FIG. 4.—The path of light rays reflected by a concave mirror, showing the centre of curvature, C, of the mirror and the focal point, D, of the reflected rays.

With a concave mirror three kinds of image can be obtained, as shown in Figs. 5, 6, and 7. In Figs. 5 and 6 the object and the image are both on the same side of the mirror, and the light rays actually pass through the image, which is called a *real* image. In Fig. 7 the object and the image are on opposite sides of the mirror, and the image is virtual, as in case of a plane mirror. Formulæ are given in textbooks which enable the exact relative positions of the object and the image to be found for various degrees of mirror curvature.

Lenses consist of pieces of special optical glass, the opposite surfaces of which have been ground to a certain shape, and they may be divided into two classes, (1) convex or converging lenses, and (2) concave or diverging lenses. In many cases lenses are not used singly, but in combinations which contain both types. A group of differently shaped lenses is shown in Fig. 8.

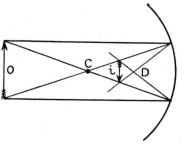

Fig. 5.—When the object, O, is beyond the centre of curvature, C, of the concave mirror the position and relative size of the image is as shown at i.

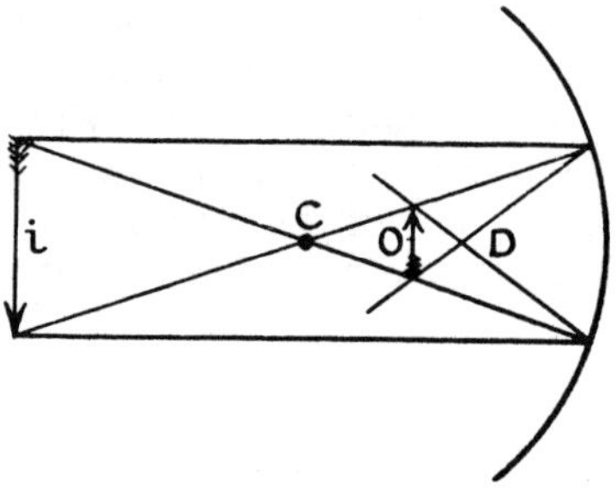

Fig. 6.—When the object is between the centre of curvature and the focal point, D, of the mirror, the position and relative size of the image is as shown at i.

The operation of a double convex lens (i.e. one which is convex on both sides) can best be explained with the aid of Fig. 9. Parallel rays of light from the object, O, instead of being reflected on striking the surface of the lens, pass in through the

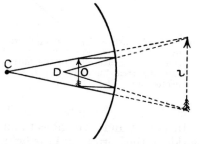

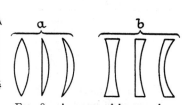

Fig. 7.—When the object is between the focal point and the mirror, the image appears *behind* the mirror, and is virtual.

Fig. 8.—A group of lenses, shown in cross-section. Those shown under *a* are convex, or converging lenses: those shown under *b* are concave or diverging lenses.

glass, but, owing to refraction, the direction of the emerging rays is changed. The emerging rays are, in fact, converged, and the point at which they intersect is called the focal point, or focus. The image of the object is formed at i. In television,

ELEMENTARY CONSIDERATIONS

double convex lenses of the type shown in Fig. 9 are most commonly employed.

With a convex lens, the image is on the opposite side of the lens so long as the object is somewhere outside the focus. The image is then real. If the object is inside the focus, then the image is on the same side of the lens, and is virtual.

In the case of a double concave lens, like that shown at the left under b in Fig. 8, the image is between the lens and the focus when the object is outside the focus, and the image is virtual. This is shown in Fig. 10.

Lenses are heir to a long list of faults, particulars of which, together with the means of guarding against them, or correcting them, will be found in any good textbook on the subject.

Another optical device which is sometimes used in television

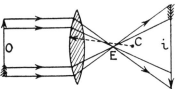

FIG. 9.—Parallel rays from the object, O, are bent or refracted by a double convex lens so that they intersect at the focus, E, and form an image of the object at i. The distance between E and the lens is called the focal length of the lens.

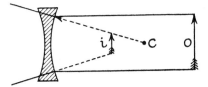

FIG. 10.—With a double concave lens, when the object, O, is outside the focus the image, i, is between the lens and the object.

apparatus is the prism. It consists simply of a triangular piece of carefully ground optical glass, and it has the property of being able to bend or reflect light. When it is used to bend light, it bends violet light most and red light least, as shown in Fig. 11. For this reason the prism is frequently used to demonstrate the visible spectrum, i.e. for breaking up what is apparently white light into the various colours which, when mixed up together, produce white light.

If a beam of light is sent into a right-angled prism in the manner shown in Fig. 12, the surface of the longest side, or hypotenuse of the prism acts as a perfect mirror, and reflects the incident beam at right angles. This property of a prism makes it invaluable for use in field glasses, range finders, and a host of other high-class optical instruments.

12 FIRST PRINCIPLES OF TELEVISION

The oldest optical device in the world is a natural one, the eye. Many of our greatest inventions are copies of something in nature, as witness, for example, the similarity between the connecting-rod of an engine and a human arm turning a mangle. Similarly, the principle of the camera is based on the principle of the human eye, and as the human eye figures so largely in television, it is as well that we should understand something about it.

A cross-section of the human eye is given in Fig. 13. It consists essentially of a double convex lens, A, and a focusing screen at the back of the eyeball, at R. The lens is situated in the front centre, or pupil of the eye, and immediately in front

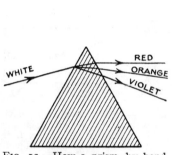

FIG. 11.—How a prism, by bending violet light more than red light, splits up white light into its component colours, producing a spectrum or "rainbow."

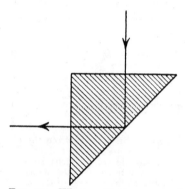

FIG. 12.—When a beam or ray of light strikes a right-angled prism as shown, it acts as a perfect reflector, bending the light ray at right angles.

of it is an iris, or diaphragm which automatically "stops down" the lens in a strong light.

In a high-class camera, using a flat plate or film, it is necessary to use an expensive combination of lenses in order to obtain an even depth of focus all over the plate, or focusing screen, otherwise the image will be sharp at the centre of the screen and blurred at the sides. In the eye, nature uses only one lens, but makes the focusing screen concave on the side facing the lens, so that the paths of rays of light between the lens and all parts of the screen are approximately equal, thus ensuring (in a perfect eye) an even depth of focus all over the screen. The lens is to some extent flexible, so that focusing for different

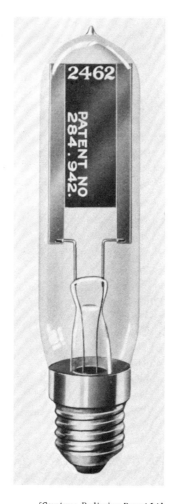

[*Courtesy: Radiovisor Parent Ltd.*

The Radiovisor Bridge, an improved form of selenium cell.

[*To face page* 12.

ELEMENTARY CONSIDERATIONS

distances can be achieved by muscular action which alters the focal length of the lens.

The focusing screen of the eye is called the retina, and its surface is composed of literally millions of hexagonal cells of microscopic dimensions. Each one of these cells is really a nerve ending, and contains a material known as visual purple. The optic nerve, O, in Fig. 13, consists in reality of a bundle of millions of tiny nerves, one for each cell. These nerves lead direct to the brain.

According to the generally accepted theory of Eldridge Green, when light falls on the visual purple in the nerve endings, a photoelectric action takes place, electrons being freed in much the same manner as they are in a photoelectric cell (see Chap. III.). These electrons set up currents which are detected near the ends of the nerves by what are called the rods and cones, which in turn set up currents which are conducted by the nerve filaments to the brain. Different intensities of light produce currents of proportionate strengths, and the brain interprets light impulses of differing wavelengths as colour. The human eye is most sensitive to light in the yellow-green region.

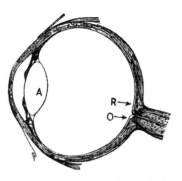

FIG. 13.—A cross-section of the human eye. The lens, A, projects an image of the scene on the retina, R, the cells of which convert light into impulses which are conveyed to the brain by the nerve channels, O.

One important feature of human vision is that the eye is able to register on the brain the entire details of a scene instantaneously. This is because, when an image is cast on the retina, the entire surface of which is covered with light-sensitive cells, these latter all respond at once, each one responding to a tiny section of the image and signalling to the brain the intensity and wavelength of the light falling on it.

A second feature of human vision which is of vital importance to television is that once an image has been impressed on the retina an appreciable period of time elapses before the impression fades. This feature is known as persistence, or retentivity of vision, and explains the phenomenon, which every reader of this book has observed at some time or another, whereby if the eyes are closed after looking intently at a bright object, an image of the object can still be seen for a period of time which depends upon the brilliance of the object, and the length of time it has been gazed at.

Both television and the cinematograph depend entirely on persistence of vision ; without this phenomenon both would be impossible. In the case of the cinematograph sixteen complete pictures (or twenty-two in the case of talking films) are flashed on to the screen per second, but owing to persistence of vision we are unable to identify them as such ; instead, we get the illusion of continuous natural movement. In television, where the same object has to be achieved, there is also presented to the observer a number of pictures in such rapid succession that the illusion of natural movement is similarly created. But in the case of television, as at present practised, persistence of vision is relied upon to a far greater extent than in the case of the cinema, as will be seen in later chapters.

The history of television may be said to date back to the accidental discovery of the light sensitive properties of the metal selenium. Before these properties were known, selenium was used in telegraphy to provide the high resistances necessary, for selenium, unlike most other metals, offers an enormous resistance to a current of electricity. Thus it came about that selenium was in use in the Atlantic cable terminal station in the village of Valentia, on the south-west coast of Ireland.

One bright sunny afternoon in 1873 the operator, named May, noticed that his instruments were behaving in a strangely erratic fashion. Investigation revealed to him that the sun, shining through the window, occasionally played upon the selenium resistances. Every time this happened, May noticed that the needle of his instrument moved.

The light-sensitive properties of selenium were thus discovered accidentally by an obscure telegraph operator, and when the discovery was communicated to the scientists of the day it created widespread interest. It opened up enormous possibilities by giving a means of turning light into electricity, and many scientists were quick to predict that selenium would provide an electric eye to supplement the electric ear which Bell's then recently invented telephone had just given them.

The earliest television experimenters attempted to produce an artificial eye by substituting selenium for visual purple and

ELEMENTARY CONSIDERATIONS

building an artificial retina out of a mosaic of selenium cells. For nerve filaments they substituted wires which connected each cell to a shutter, which took the place of the brain. For every selenium cell there was a shutter, and each shutter was arranged to open when light fell upon the particular cell connected to it. As each shutter opened, it allowed a spot of light to fall upon a screen at the receiving end of the circuit. In this way each selenium cell controlled a spot of light, the image being produced by a mosaic formed of these spots.

Apparatus modelled on these lines was actually made by several inventors. Rignoux and Fournier, two French scientists, constructed such a machine in 1906. This apparatus was intended only to demonstrate the principle, and made no pre-

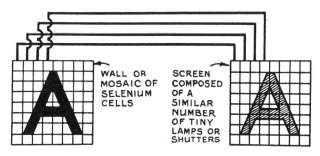

FIG. 14.—Showing diagrammatically Rignoux and Fournier's apparatus.

tensions towards being an instrument for the actual accomplishment of television.

The transmitter consisted of a wall covered with selenium cells, sixty-four fairly large cells being used (see Fig. 14). From each of these cells two wires ran to the receiving screen, which was made up of sixty-fours shutters, each shutter being controlled by its respective selenium cell. Thus, when any given cell at the transmitting station was brilliantly illuminated, it sent a strong current over the two connecting wires to the receiving station, where it caused the appropriate shutter to open so that light from a small lamp could shine through on to the corresponding part of the receiving screen. By covering the transmitting wall with large stencils, and shining a powerful light on the wall, images of letters of the alphabet were transmitted and could be recognised at the receiver.

16   FIRST PRINCIPLES OF TELEVISION

Ernest Ruhmer, whose brilliant pioneer work in connection with wireless telephony is so well known, also constructed a similar apparatus, and many other workers have been attracted to this system. However, the thousands of cells, shutters, and connecting wires necessary made the practical adaptation of such schemes entirely out of the question, and an endeavour was made to solve the problem in quite a different manner. In passing, however, it is interesting to note that certain receiving apparatus which has been used recently is strangely reminiscent of this early scheme. This will be described in a later chapter.

The suggested alternative was to divide the scene up into a great number of small parts and then transmit it section by section to the receiving end, where the scene was to be reassembled again like the pieces of a jig-saw puzzle. The entire operation was to be carried out so rapidly that the effect upon the observer would be the same as that of an instantaneous glance by direct vision.

Imagine, for example, that by means of a lens an image of the object or scene to be transmitted were to be thrown upon a ground glass screen, just like the focusing screen of a camera. Imagine this image divided up into thousands of little squares like a chessboard, each one of these little squares being dark or light, depending upon what part of the picture it belonged to.

Now we have to send this picture over a wire or wireless circuit. For the sake of an example let us assume that there are 1000 squares numbered 1, 2, 3, 4, 5, 6, 7, etc., up to 1000 ; then if we send a series of messages, 1 dark, 2 light, 3 light, 4 dark, etc., to an assistant at the receiver who has before him a board also divided into 1000 squares, and if he makes his squares light or dark as directed, he will build up a mosaic similar in all respects to that on the ground glass screen at the transmitting end.

Now, in this operation, we looked at each little square at the transmitting end, and our eyes told us whether it was dark or light. If, instead of using our eyes, we passed a selenium cell over the squares, it would send out a series of electrical impulses which would be strong or weak in accordance with the lightness or darkness of the squares.

If, at the receiving station, we had, instead of a human assistant, a mechanical device which would direct a narrow beam of light from a lamp on to each square of the receiving board in

ELEMENTARY CONSIDERATIONS

turn, at exactly the same time as the cell at the transmitting station passed over the corresponding square on the ground glass screen ; and if, further, the intensity of the light was arranged to be controlled by the varying current from the cell, then each square of the board at the receiver would in turn be illuminated with a brightness corresponding to the brightness of the same square on the transmitting screen.

If this entire operation were to be carried out in not more than one-eighth of a second, then the observer would see on the receiving board, not a moving light-spot of varying intensity, but the whole image at once, thus giving the effect of an instantaneous glance. In other words, use would have been made of persistence of vision to mask the movements of the light-spot.

Upon the principle of exploring an image described above, a great number of television systems have been evolved, many unsuccessful, but it is to development along these lines that success was ultimately due.

CHAPTER III.

LIGHT-SENSITIVE DEVICES.

Simple Selenium Cell. Nature of Selenium. Time Lag. Radiovisor Bridge. Photoelectric Effect. Vacuum and Gas-filled Photo Cells. Amplification. Zworykin's Combination Cell. Shadowgraphs and Television Differentiated. Requirements for Television.

SELENIUM cells were briefly referred to in the last chapter. Since the light-sensitive device, of whatever type it may be, is an essential component of every television transmitter, whatever system is employed, the time has come to review all the principal devices which are capable of transforming light energy into electrical energy.

All light-sensitive devices depend for their action upon what is known as the " photo-electric phenomenon." This phenomenon has been subdivided into two distinct types, photo-conductivity and photo-emissivity.

The first of these was originally discovered, as described in the last chapter, by a telegraph operator who noticed that when the sun shone on some selenium resistances the needle of his instrument moved. This discovery led to a detailed investigation of the phenomenon by the leading scientists of the day, notably Willoughby Smith, with the result that the fact was established that so long as selenium is kept in darkness its resistance to electricity is extremely high. As soon as it is exposed to light, however, its resistance drops very considerably.

This fact established, efforts were immediately made to make use of selenium as a means of transforming light into electricity, and for this purpose special devices, known as selenium cells, were constructed, a simple method of manufacture being as follows :—

Two wires, spaced about ·5 mm. apart, are wound simultaneously over a flat piece of insulating material, such as steatite. These wires do not make contact with each other at any point. A coating of amorphous selenium is then spread over the wires, filling the spaces between them, and a heat treatment is then applied which has the effect of turning the selenium into the grey crystalline form which has been found to be most sensitive to light variations.

The method of construction is shown in Fig. 15.

An average cell made up in this manner has an electrical resistance in darkness (between the ends of the two wires) of between 60,000 and 100,000 ohms. By illuminating the cell with a 16-c.p. incandescent lamp placed at a distance of 1 metre, its resistance drops to approximately 30,000 or 40,000 ohms, or about half the resistance of the cell when it is in darkness.

Selenium itself is a rather curious substance, belonging

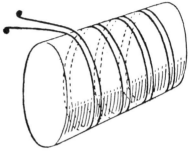

FIG. 15.—Showing the construction of a simple form of selenium cell.

to the sixth group in the periodic classification of the elements. It takes its place between sulphur and tellurium, which explains most of its remarkable properties. It lies just between the metals and the non-metals, and shows the tendency of sixth-group elements to form allotropic modifications; thus it exists in the following three forms: (1) Amorphous selenium (including the vitreous and colloidal forms); (2) Crystalline red selenium, which exists in two forms; (3) Crystalline grey metallic selenium. It is only in the last form that selenium is conductive and light sensitive.

Other substances which exhibit photoconductivity (but in a lesser degree) are antimonite, a natural sulphide of antimony, thallium oxysulphide, and sulphur.

The theory of the action of photoconductivity is that, under the action of light, electrons are instantly liberated which are capable of free motion within the metal when an electromotive force is applied. When the light is withdrawn, these electrons are recombined again, but the recombination process takes a certain period of time which is proportional to the square of the number of electrons present, thus giving rise to the well-known time lag of the selenium cell.

It is this time lag which has so far rendered selenium cells totally unsuitable for television work, and it was not until the development of the photoelectric cell that television became possible, except in a crude form. For television purposes a light-sensitive device is required which will not only respond strictly proportionately to the light intensities falling upon it, but will also respond instantaneously without any measurable lag.

A curious feature of selenium is that it responds much more efficiently to light of weak intensity. This is because, for strong light, the response is proportional to the square root of the illumination, whereas for weak light the response is directly

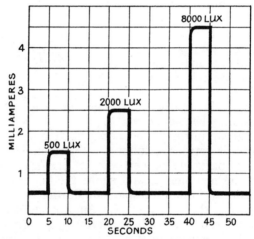

Fig. 16.—Curves showing the response of selenium under strong illumination. The current is proportional to the square root of the illumination.

proportional to the illumination. Figs. 16 and 17 illustrate diagrammatically the response of selenium to strong and weak illumination.

Under feeble illumination, the output of selenium is remarkable. According to Dr. E. E. Fournier D'Albe, the well-known authority on the subject, if it gives a current of 1 milliampere under an illumination of 100 lux (one lux is one candle-power at a distance of one metre), it will still give a current of 100 microamperes under an illumination of 1 lux, and 1 microampere under an illumination of 1/10,000 of a lux.

There has recently been developed a new and highly efficient type of selenium cell, known as the Radiovisor Bridge.

LIGHT-SENSITIVE DEVICES

The ordinary type of selenium cell will not, as a rule, stand an energising potential much greater than 50 volts, but the Radiovisor Bridge will stand as much as 1000 volts for short periods, and several hundred volts for months on end without suffering deterioration. The device is being used in England for automatically controlling street lighting, and for burglar alarms. More recently, in an improved form, it has been successfully used in talking film projectors instead of a photoelectric cell.

The construction of the cell, or bridge, is as follows. A gold grid is first fused on to the surface of a thin sheet of glass by a

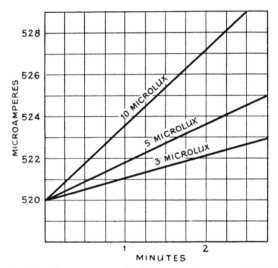

FIG. 17.—Showing the response of selenium under very faint illumination. The current is directly proportional to the illumination.

special ceramic process. The grid takes the form of a pair of interdigitated combs which form the electrodes (see Fig. 18). As the gold layer is very thin and the combs only approximate at their edges, the capacity is very small. This is an important point if the cell is to be used for certain purposes. Molten selenium is spread very thinly over the surface, and is then subjected to a very carefully controlled thermal process, as a result of which the selenium is converted into the crystalline light-sensitive variety. The selenium layer is of the order of $2 \cdot 5 \times 10^{-3}$ cm. in thickness, and the effect of such a thin layer on a transparent base is to make the utmost quantity of the material accessible to light, and to leave as little as possible to act as an inert shunt to the active portion.

The actual area of the sensitive surface varies considerably in size for various purposes, but the surface of the standard cell measures 25 mm. by 50 mm. When complete the cell is mounted on German silver clips which make contact with the grid electrodes, and enclosed in a glass bulb which is first evacuated and then filled with a chemically inert gas. The bulb is then fitted with a standard screw socket (the centre contact of which is the positive electrode) and subjected to an ageing process. A

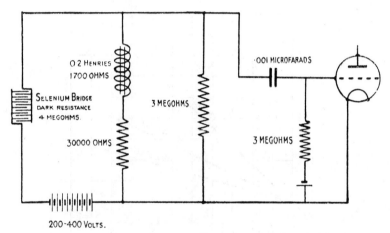

FIG. 18.—A circuit for use when the Radiovisor Selenium Bridge is to be employed for acoustic work. The E.M.F. produced across the inductance and resistance is applied to an amplifier, the first valve of which is shown.

standard cell of about 4 megohms will give a change between current in darkness and in light of the order of 100 to 150 microamperes.

Owing to the special process of manufacture the Radiovisor Bridge will respond to rapid fluctuations of light much better than an ordinary selenium cell. Measurements have shown, in fact, that it will respond to light variations occurring at frequencies up to 8000 per second. It is therefore suitable for such acoustical applications as talking films, where it has the advantage over the photoelectric cell that it requires less amplification, thus reducing the complications of extensive

LIGHT-SENSITIVE DEVICES

amplification and the " ground noise " which such a high degree of amplification inevitably produces.

Selenium is not equally sensitive to all colours of light. It gives by far the greatest response in the orange region of the spectrum.

Owing to the high coefficient of resistance of selenium, care must be taken that the energy dissipated in the cell does not cause an excessive rise of temperature. For this reason it is usual to insert a high resistance in series between the cell and the energising electromotive force. Fig. 18 shows a circuit which is suitable for use in acoustic work, the inductance and resistance being inserted in series to prevent the over-accentuation of the lower frequencies. The grid of the first valve of an amplifier is shown connected across this combination.

Thus far has the selenium cell progressed in recent years. It remains to be seen whether still further improved methods of construction will reduce the lag even more so as to enable it to respond to the high frequencies encountered in television work.

The phenomenon of photo-emissivity was first discovered by Hertz in 1888. He observed that when violet light fell upon one of the electrodes of a spark gap the electric discharge passed more readily, i.e. the resistance of the gap was reduced and a longer spark resulted. Hertz at that time was concentrating on his classic experiments with electromagnetic or Hertzian waves, out of which arose wireless telegraphy, and beyond noting the effect he paid but little attention to it. The phenomenon was investigated by another German scientist named Hallwachs, who discovered that the effect was located at the negative pole of the spark gap.

Further investigations by subsequent workers disclosed the fact that all metals, under the influence of light, emit electrons, but the effect is much more pronounced in the case of metals belonging to the alkali group, such as rubidium, potassium, sodium, and cæsium. The following table (prepared by the G.E.C.) gives an idea of the current response obtained per lumen of light from cathodes of different metals, using as light sources gas-filled lamps and the sun. It is interesting to note the immense superiority of sunlight over artificial light. This is because sunlight is rich in violet rays, to which the metals are chiefly responsive. Response curves which reveal the characteristics graphically are given in Fig. 19.

Metal.	Emission.	
	Gas-filled Lamp. Amp. lumen.	Sunlight. Amp. lumen.
Na	$\cdot 2 \times 10^{-6}$	2×10^{-6}
K	1·0	6·4
Rb	·4	2·1
Cs	·15	·4
Kon Cu	·8	1·7

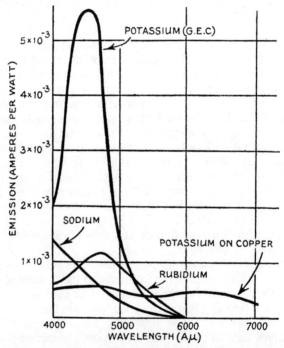

Fig. 19.—Showing graphically the degree of photoelectric response of different metals when exposed to light. The response is greatest for light in the violet region of the spectrum.

By making use of the photoelectric effect a new type of light-sensitive cell, known as a photoelectric cell, was developed. This type of cell consists essentially of a plate of one of the alkali metals, enclosed in a high vacuum and connected to the negative

[*Courtesy: Westinghouse Electric & Manufacturing Co.*

Dr. V. Zworykin and his combination photoelectric cell and thermionic valve.

[*To face page* 24.

LIGHT-SENSITIVE DEVICES

terminal of a battery. The plate forms the cathode of the cell, the anode (to which the positive terminal of the battery is connected) usually consisting of a loop of wire or a wire gauze mounted opposite the cathode. The potential used to energise the cell is usually of the order of 200 volts.

The pioneers in the application of this type of cell to the

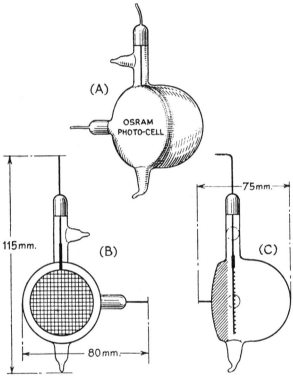

FIG. 20.—The above sketches show the external and internal arrangement and appearance of the standard G.E.C. photoelectric cell.

measurement of visible light were Elster and Geitel. By 1905 they had established all the main principles, apart from the amplification of the output by means of thermionic valves, which were not then available. As a result of the impossibility of magnifying the output, little use was made of photoelectric cells for many years until modern high efficiency valve amplifiers began to be developed.

Not all photoelectric cells are enclosed in a vacuum. In some cases inert gases at low pressure are introduced into the bulb, and when using such cells care must be taken not to use too high an energising potential, or a glow discharge will be produced within the cell which will ruin it if maintained.

Fig. 20 shows sketches of the standard G.E.C. Osram gas-

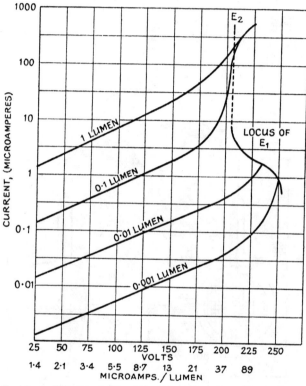

FIG. 21.—Characteristic curves of the standard G.E.C. photo cell.

filled cell, which has a cathode of potassium (thinly deposited on the inside of part of the glass bulb) specially sensitised by an electric discharge through hydrogen. The cell is filled with argon to a pressure of about 0·15 mm. Characteristic curves of the cell are given in Fig. 21.

An elementary circuit arrangement, explaining the action of a photoelectric cell, is given in Fig. 22.

LIGHT-SENSITIVE DEVICES

There are essential differences between vacuum and gas-filled photo cells. In a vacuum cell the output current is due solely to the electrons liberated from the cathode when the latter is exposed to light. In a gas-filled cell the action is more complicated, due to the ionisation which takes place, and the output current is much greater. Electrons liberated from the cathode, in the course of their journey to the anode, encounter molecules of gas with which they collide. Each molecule of gas consists of a positive nucleus around which revolves a planetary system of negative electrons. The result of every collision is that one of the planetary electrons gets knocked out of its orbit and proceeds to travel independently towards the anode, leaving the gas molecule minus one of its electrons.

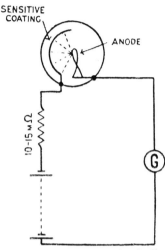

FIG. 22.—Elementary circuit diagram of a photo cell and associated external circuit. When light falls on the cell a current flow will be indicated by the sensitive galvanometer, G.

We now have two electrons, instead of one, making for the anode, and both may have further collisions before reaching their destination. The net result, therefore, is that the original or primary electron stream given off by the cathode becomes greatly amplified before it reaches the anode, and this amplification has the effect of increasing greatly the output current of the cell.

Vacuum cells are preferable to gas-filled cells in everything but sensitivity. If accuracy is required, as in scientific research work, efforts should be made to use lights strong enough to give the requisite current in a vacuum cell, and to use the current to the best advantage, so that only a small current is required. But for most industrial purposes, and especially in television work, where sensitivity rather than accuracy is the first requisite owing to the very low light intensities available, gas-filled cells are necessary. The choice of the nature and pressure of the gas requires care; but this is a problem for the manufacturer, not the user.

As we have already seen, photoelectric cells are chiefly sensitive to light rays belonging to the violet end of the spectrum, and as the red end is approached their sensitivity falls off. However, the needs of television have caused a vast amount of research work to be carried out in connection with photo cells, and such rapid progress in their design and manufacture has been made in the last few years that we may expect to hear at any time that a panchromatic cell has been developed.[1]

In contrast to photoelectric cells, selenium is very responsive to orange and red light, and comparatively insensitive to rays belonging to the upper end of the spectrum. Selenium also gives a vastly greater current output under the influence of light than does any known form of photoelectric cell. Whereas the latter, when illuminated, will give a current of only a few microamperes, the ordinary selenium cell will give a current which can be measured in milliamperes. Special large selenium cells have actually been constructed which will give a current of 1 ampere, such enormous cells being used in the Cox Selenium Relay, employed in cable work.

As already explained, photoelectric cells depend for their action upon photo-emissivity. That is to say, under the influence of light the cathode emits electrons. Since the velocity of electrons is of the same order as that of light, i.e. about 186,000 miles per second, it follows that the response of the photoelectric cell will be, to all intents and purposes, instantaneous. This proves to be the case, and as soon as photoelectric cells made their appearance, television workers who had been baulked by the lag of selenium immediately seized upon them in the belief that their difficulties were ended.

However, although photoelectric cells have absolutely no time lag, they experienced nothing but disappointment because, as we have seen, the output of the vacuum cells is of the order of a few microamperes only, and gas-filled cells had not been developed sufficiently. Such a minute current can be detected in the laboratory by means of a sensitive galvanometer connected as shown in Fig. 22, but before it can do any useful work it must be magnified very considerably by means of thermionic valve amplifiers. It is only within the last five years

---

[1] It is not possible within the scope of this book to go too deeply into the subject of photoelectric cells. Readers who are interested to study the subject more closely are referred to a recently published book entitled " Photoelectric Cells," by Dr. Norman R. Campbell and Dorothy Ritchie (Sir Isaac Pitman & Sons, Ltd.).

or so that amplifiers have been developed which are capable of giving the degree of amplification necessary for television purposes. Prior to this it was found that when the amplification was pushed beyond a certain point, parasitic noises were generated, due to the irregular emission of valve filaments and irregularities in the associated batteries, and to tendencies towards self-oscillation in the amplifier circuits. These difficulties prevented a sufficiently high level of amplification of the photo-electric currents being obtained.

Fig. 23 illustrates the fundamental amplifier circuit for use with photoelectric cells; from this circuit all variants are derived.

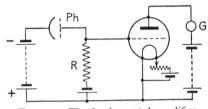

FIG. 23.—The fundamental amplifier circuit for photoelectric cells.

A resistance-capacity coupled amplifier circuit (to which further stages can be added) is shown in Fig. 24. In this circuit the choice of the resistance R_1 is the only matter requiring special consideration. The greater R_1 the greater will be the variation of the grid potential of the first valve produced by a given photoelectric current, and the greater, therefore, the sensitivity. But a limit is set by distortion of the higher frequencies. The cell itself

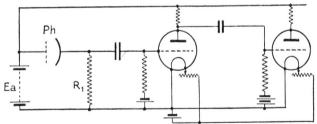

FIG. 24.—A resistance-capacity coupled amplifier circuit for photoelectric cells.

possesses a definite capacity, and in order to combine as far as possible sensitivity and absence of distortion, this capacity must be made as small as possible and, further, the cell should be placed as close as possible to the first valve, in order that the capacity of the leads may be reduced to a minimum. Even then R_1 should not usually exceed 2 megohms.

In an effort to overcome the low output handicap of the

photoelectric cell, V. K. Zworykin, of the Research Department of the Westinghouse Electric Company (U.S.A.), devised in 1926 an interesting form of cell which incorporates within the same glass bulb the elements of a three-electrode valve.

The lower portion of the bulb contains the valve elements, whilst the upper half, which is screened from the valve portion by a light-tight partition, contains the photoelectric cell. The light-sensitive substance used in the cell is potassium hydroxide, and this is spread in the form of a coating over one inner side of the upper part of the glass bulb.

The valve filament is of the oxide coated variety, which operates without perceptible glow. By using such a filament, no unwanted light from this source can reach the light-sensitive coating of the cell. A grave disadvantage, however, is that alkaline photoelectric substances are highly volatile, and the heat from the filament in such close proximity is liable to cause damage to the cell.

However, using such a combination, Zworykin claimed to be able to obtain an amplified output of 1 milliampere. For higher outputs a type of 4-electrode valve is employed, using the first grid as an anode with a low potential of the order of about 30 volts positive. The second grid is connected in orthodox fashion, and the plate has a positive potential of the order of 150 volts impressed upon it. By a proper choice of spacing between electrodes, and grid mesh, it is possible to obtain a good relationship between the degree of illumination and current output within certain limits. Using the 4-electrode valve combination, Zworykin claimed that a continuous current output of 5 milliamperes has been obtained, the limiting factor being the heat developed within the photo cell, which causes distillation of the metallic coating, with the result that it condenses on the transparent glass and insulating parts. It has been stated, however, that this difficulty can be remedied to a great extent by improved construction.

Summing up the conclusions arrived at in this chapter, we find that selenium, while sensitive enough for television purposes, is far too slow in its response. The photoelectric cell, on the other hand, is instantaneous in its response but lacking in sensitivity, which disadvantage rendered it also unsuitable for television until modern high efficiency amplifiers were perfected.

By using the earlier forms of photoelectric cell and amplifier, however, many workers succeeded in transmitting shadow-

A close-up view of the Belin apparatus. Finger is pointing to the tiny vibrating mirrors.

[*To face page* 30.

graphs, or silhouettes of objects placed before the transmitter. In this case the question of the sensitivity of the cell is not of much importance, for it can be directly illuminated by a beam of light which may be as strong as we care to make it. As the object the shadowgraph of which is to be transmitted is interposed between the source of light and the photoelectric cell, all the latter has to do is to distinguish between a strong direct light and no light at all, i.e. a shadow of the object.

When transmitting shadowgraphs the illumination of the cell by the light source is *direct;* in television it is *indirect.* That is to say, when true television is to be accomplished, the source of light must be arranged to shine directly on to the object to be transmitted, not the cell. Then the amount of light

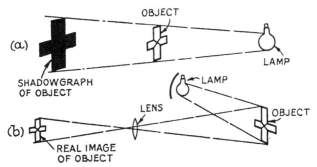

FIG. 25.—By using direct lighting, a shadowgraph of the object is obtained, as in (*a*). By using indirect or reflected light, as in (*b*), a real image of the object is obtained, showing detail.

which reaches the photoelectric cell is that which is *reflected back* from the object on to the cell. This distinction between direct and reflected light is illustrated in Fig. 25.

Reverting for a moment to natural vision, all self-luminous bodies, such as the sun, are seen by direct lighting. All non-self-luminous bodies are seen by indirect lighting, i.e. light from the sun strikes the bodies, which reflect some of the sun's rays into our eyes.

To those unacquainted with the subject there may not appear to be any particular significance in the above explanations, but the important point will be grasped when it is remembered that in ordinary vision only a minute fraction of the sun's light is reflected back into our eyes by the objects we most commonly look at. Compare, for instance, the light reflected from a piece of wood with that reflected from a mirror set at the right angle.

When it comes to television, using artificial light, the *total* amount of light reflected from an object such as the human face, when it is illuminated by lamps of one thousand candle-power, at a distance of one metre, is less than that of a single candle. Furthermore, owing to the process of image analysis or scanning which occurs in television, whereby the image is subdivided into thousands of tiny pieces or elements, only an infinitesimal fraction of this already very small amount of light is available to act on the cell at any given instant.

The ideal light-sensitive device for television purposes, therefore, must combine a high degree of sensitivity with instantaneity of response.

CHAPTER IV.

SOME EARLY TELEVISION EXPERIMENTS.

Szczepanik's Vibrating Mirror System. Rosing's Mirror Drums and Braun Tube. Cathode Rays. Mihaly's Oscillographs. Belin and Holweck's Cathode Ray Oscillograph.

HAVING examined the elementary requirements for the accomplishment of television, together with some of the essential apparatus, we can now proceed to a study of some of the complete systems which have been devised from time to time. It is not necessary to weary the reader with detailed descriptions of *all* the apparatus which has ever been experimented with; it is only necessary, in dealing with the earlier attempts to solve the problem of television, to give the details of a few of the machines in order to give the reader an adequate idea of the historical background. In succeeding chapters we will go further into the details of apparatus of more modern design which has actually proved successful in operation.

It is, however, necessary to pay tribute to the efforts of the early workers who, although unable themselves to achieve success, laid the foundations upon which, later on, the inventive genius of their successors, aided by the availability of more modern equipment, was to build.

Fig. 26 illustrates diagrammatically one of the most representative and suggestive of the early television schemes, due to Jan van Szczepanik.

Referring to the diagram, the light rays reflected from the scene to be transmitted, X, are focused by means of a lens firstly on to the mirror, A, which is maintained in a state of rapid vibration by the electromagnet, L. From A the beam of reflected light is reflected again on to the mirror, B, which is kept vibrating at a much slower rate by the electromagnet, J. This second mirror vibrates in a plane at right angles to the first one. The combined action of these two mirrors causes the beam of light containing the image to traverse the aperture leading to the light sensitive cell, C, in a zig-zag path.

In order to make the action clear, let us consider the action of the camera. If the lens is opened, and a ground-glass screen

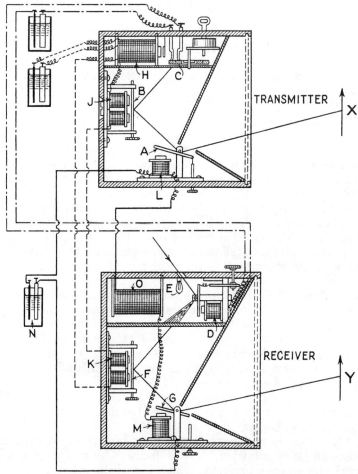

FIG. 26.—Jan van Szczepanik's television apparatus.

put in position instead of the sensitive plate or film, an image of the scene before the camera will appear on the screen. In the television transmitter, an image of the scene, X, appears in exactly the same manner on the first mirror, A.

SOME EARLY TELEVISION EXPERIMENTS 35

The combined action of the two vibrating mirrors is the same as if we could, in some way, take the focusing screen of the camera, with the image of the scene still showing on it, and rapidly move it to and fro in a zig-zag path across the aperture of the light-sensitive cell, C. If we did this, it is clear that eventually the aperture of the light-sensitive cell would have looked at, scanned, or explored the entire surface of the ground glass screen. That is to say, it would have scanned the entire image visible thereon, and been influenced by the degree of lightness or darkness of parts of the image which passed over it.

At the receiver the operation is exactly similar to that of the transmitter, only the light-sensitive cell is replaced by the lamp, E. A beam of light from this lamp is first focused on to a vibrating mirror which is controlled by the magnet, D, which, in turn, takes its supply current from the electrical impulses sent to the receiver by the light-sensitive cell at the transmitter. Thus the intensity of light reaching the mirror, F, depends upon the extent to which the light beam is deflected away from a small aperture in the partition between D and F, and this deflection, as explained above, depends upon the strength of the currents arriving from the transmitter.

This varying light beam is then reflected on to the vibrating mirror, F, and from thence to the third vibrating mirror, G. As in the case of the transmitter, these two latter mirrors vibrate in planes at right angles to one another, thus giving to the light beam a double movement; one a rapid to-and-fro motion; the other a more gradual movement downwards, imparted by the mirror which vibrates at the slower speed.

The resultant action of the apparatus is that the beam of light is spread over the screen, Y, moving rapidly to and fro across it, and slowly down from top to bottom. Since the light beam is varying in intensity, due to the action of the magnet, D, it follows that the intensity of illumination of different parts of the screen will be different, corresponding exactly with the light and shade of the scene at the transmitter. The entire action takes place so rapidly that, to the eye of an observer, the screen appears to be continuously illuminated with the image being received.

That is the general arrangement of Szczepanik's apparatus, but successful results were never achieved with it. Two insurmountable obstacles to success remained. One of these was that a suitable light-sensitive device was not available;

we have already seen the shortcomings of the selenium cell. The other was the difficulty of maintaining synchronism, i.e. arranging for the mirrors at both transmitter and receiver to vibrate exactly in synchronism, or in step with each other. Unless the mirror systems at both ends of the circuit operate at exactly the same speed, successful results cannot be obtained, for the received picture becomes hopelessly muddled up.

Another scheme of a more interesting character is that suggested in 1907 by Boris Rosing, a Russian professor. His transmitting arrangement is similar in many respects to others, but his receiving device is very original, in that it dispenses altogether with mechanical parts and uses, instead, a cathode ray or Braun tube. The particular point of interest is that apparatus, both transmitting and receiving, similar to that proposed by Rosing is still being experimented with to-day by a number of workers.

The cathode ray is a form of electrical discharge which occurs when electricity at very high potential is forced through a very high vacuum. The rays can be produced in the form of a thin pencil-like stream, and if directed on to a special form of fluorescent screen, they can be rendered visible.

A cathode ray stream has the important characteristic, from a television point of view, that it is capable of being moved, or diverted from its path, in any direction, either electrically or magnetically. Since a cathode ray stream consists really of a stream of electrons, it has no weight, and therefore no inertia, and there is thus no limit to the speed at which it can be caused to move about.

When the rays strike a plate of flourescent material, a brilliant spot of light is produced, so that by using cathode rays in conjunction with a fluorescent screen we have the elements of a receiving device which is capable of responding instantly at practically any speed.

Rosing's arrangement is shown in Fig. 27. At his transmitter he used two mirror polyhedrons revolving at right angles to each other, as shown in the diagram at C and D. As in the case of Szczepanik's vibrating mirrors, one mirror polyhedron revolves much more slowly than the other. The combined motion of these mirrors causes an image of the object to be transmitted, A (focused on to C through the lens B), to be swept across the aperture of the selenium cell, H. The action of the image-scanning mechanism is in all respects similar

[*Photo: Harris & Ewing.*

C. Francis Jenkins photographed with (left) a lens disc and (right) his prismatic disc.

SOME EARLY TELEVISION EXPERIMENTS

to that described previously in connection with Szczepanik's apparatus.

The varying current impulses from the cell, H, are transmitted to the receiver, where they are caused to charge two condenser plates, L, in the cathode ray tube. The fluctuating charge impressed on these condenser plates produces electrostatic stresses which cause the cathode ray beam to be deflected away from the aperture, M, placed in its path, and the amount of the ray which passes through the aperture is thus made proportional to the potential of the plates, i.e. to the degree of light and shade in the original scene.

Having got through the aperture, M, the ray then strikes

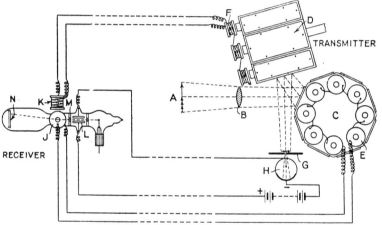

FIG. 27.—Boris Rosing's television apparatus, employing a cathode ray tube.

the fluorescent screen, N. The manner in which the ray is caused to bend to and fro, thus covering the entire screen, is quite ingenious.

The revolving mirror polyhedrons carry magnets which revolve with them. These magnets act upon groups of fixed coils, set close to the polyhedrons, setting up alternating currents in them. Two alternating currents are generated in this way, one of them by one revolving polyhedron, and the other by the second polyhedron. Both currents are sent along separate pairs of wires to the receiver, where they are arranged to energise two electromagnets, K and J, one of which gives to the cathode ray vertical movement, while the other gives to it a much more rapid horizontal motion.

FIRST PRINCIPLES OF TELEVISION

Successful results were not obtained with this apparatus, due partly to the selenium cell difficulty, but more particularly to difficulties in connection with the crude form of cathode ray tube then available.

The idea of using cathode rays for television purposes, both at the transmitter and at the receiver, has attracted the attention of quite a number of workers, including contemporary experimenters. The advocates of the cathode ray argue that television in its widest sense will never be achieved by the mechanical methods which are at present being used with a fair measure of success. It is contended that the inertia, wear and tear, and other limitations of mechanical devices preclude further advance; and that, for all mechanical devices at present in use,

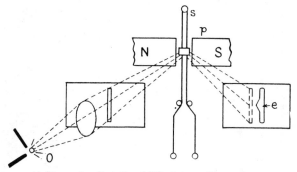

FIG. 28.—Details of Mihaly's oscillograph.

purely electrical means will have to be substituted, because electrons, being weightless, have no inertia and are capable of being directed at any speed. The advocates of this line of argument therefore turn to the cathode ray tube, the only device at present known which conforms to their requirements.

However, as at present constructed and used, cathode ray tubes are very expensive and variable in performance, their average life is only about 200 hours, and the associated equipment is both expensive and complicated. There is also the difficulty of focusing the rays to a sufficiently fine point to prevent " spread " and consequent loss of detail in the received image.

We shall return to a consideration of modern cathode ray systems in later chapters.

In more recent years Denes von Mihaly, of Budapest, has

SOME EARLY TELEVISION EXPERIMENTS

experimented with an apparatus which, although very complex, is nevertheless of interest.

The principal feature of this apparatus is the image-scanning mechanism. This takes the form of a system of very small oscillating mirrors, having an area of one square millimetre or less. The arrangement is shown in detail in Fig. 28, where the mirror, P, is fixed to a loop of extremely fine platinum wire, S. The diameter of this supporting wire is of the order of ·01 mm., and it is stretched between the poles of a powerful electromagnet, NS. This arrangement is a form of the well-known Siemens oscillograph.

The transmitting apparatus is shown diagrammatically in Fig. 29. The lenses, a and b, reduce in area the image of the

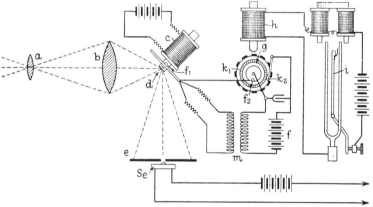

FIG. 29.—Mihaly's early transmitting apparatus. The scene to be transmitted is placed to the left of the lens, a.

scene which is to be transmitted, and project it on to the small oscillograph mirror, d. This mirror oscillates at a frequency of 500 vibrations per second, and is kept oscillating at this speed by means of a 500-cycle alternating current which is caused to flow through the fine platinum suspension wire.

The same mirror also vibrates at a much slower frequency in a plane at right angles to that of the 500-cycle oscillation, the slower vibration being imparted to it by means of the lever connecting it to the phonic wheel, g. This wheel, or drum, is driven by the electromagnet, h, which, in turn, receives its alternating current supply from the tuning-fork, i. This tuning-fork vibrates at a frequency of 100 cycles per second, and the drum, *g*, which it is responsible for driving, rotates at a speed of 300 revolutions per minute.

The revolving drum and tuning-fork form part of the synchronising apparatus, further details of which will be given in the chapter devoted to synchronism.

The manner in which the image of the scene to be transmitted is broken up into its component picture elements is as follows. The reduced image projected on to the oscillating mirror is reflected therefrom on to the screen, *e*. This reflection is approximately the same size as the original image, on account of the " swing," or divergence of the light rays which results from the oscillations of the mirror.

In the middle of the screen, *e*, there is an aperture approximately 1 millimetre wide, which runs the full length of the screen, or diaphragm. Behind this aperture is placed a selenium cell, and the action of the mirror causes the complete image of the scene to traverse the cell in one-tenth of a second.

As already stated, the drum, revolving at 300 revolutions per minute, oscillates the mirror in one plane at the rate of $300 \div 60 = 5$ vibrations per second. It is oscillating in a plane at right angles to this motion at 500 vibrations per second.

Therefore, in one-tenth of a second, while the mirror makes 50 oscillations in a horizontal plane, half an oscillation is made in a vertical plane. This means that the image is swept across the aperture leading to the selenium cell in 50 horizontal lines, which are drawn across the screen, *e*, one below the other. The 1-millimetre aperture in the screen ensures that each line, instead of striking the selenium cell all at once, does so from one end to the other, in 1-millimetre graduations.

The current from the selenium cell, Se, is amplified and sent to the receiving station.

The receiving equipment, shown in Fig. 30, is somewhat similar in arrangement to the transmitter. The same forms of tuning-fork, *i*, phonic drum, *g*, and oscillograph mirror, *d*, are used. Instead of a selenium cell, there is in the receiver a so-called "light relay," which is a device which converts the incoming picture current fluctuations into light fluctuations, which are spread over the receiving screen, *t*.

The light relay consists of a very sensitive bifilar oscillograph of special design. The arc lamp, *o*, through a system of lenses, projects a narrow but extremely intense beam of light on to the mirror, p, of the oscillograph. The fluctuating picture current from the transmitter is conducted through the two fine wires supporting the mirror, and as these wires are in the field of the electromagnet, q, the mirror is deflected in direct proportion to the strength of the current flowing through its supporting wires.

This deflection causes more or less of the light beam to fall on the aperture, e, all the beam passing through it when the received current is strong (indicating a brilliantly illuminated section of the scene at the transmitter), and only a very small

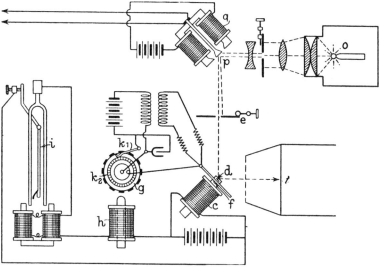

FIG. 30.—Mihaly's early receiving apparatus.

portion of it, or none at all, when the received current is weak, or nil.

When no current is passing through the suspension loop of the mirror, p, the narrow beam of light reflected from the mirror is projected close to the diaphragm aperture, e, but no light passes through it.

As soon as a current flows, causing a deflection of the mirror, some of the light beam passes through the aperture and falls on to the second oscillograph mirror, d. This mirror, through the agency of the tuning-fork-phonic-wheel combination, and a 500-cycle alternating current, vibrates in exactly the same manner, and in step with its counterpart at the

42 FIRST PRINCIPLES OF TELEVISION

transmitter. Thus, by means of the reflected light falling upon the vibrating mirror and being reflected on to the screen, t, the picture is reproduced.

In spite of the ingenuity of the above arrangement, television could not be achieved, due, partly, to the unsuitablity of the selenium cell, but principally to the fact that the optical system is incorrect. Mihaly described his oscillograph mirror as being " a thin flat mirror." Referring to Fig. 29, if the mirror, d, has a plane surface, no image will be reflected on to the screen, e, but only a blur. Difficulty was also experienced in keeping all the oscillograph mirrors vibrating in synchronism. By means of this apparatus, however, Mihaly succeeded in transmitting and receiving crude shadowgraphs. Details of his present-day equipment will be given in a later chapter.

In France, shadowgraphs of simple objects have been transmitted by M. Belin (whose name is well known in connection with phototelegraphy), working in conjunction with M. Holweck of the Radium Institute, Paris. A diagrammatic outline of the apparatus used is given in Fig. 31, from which it will be seen that the arrangement of the transmitter is not unlike that of the systems of Szczepanik and Mihaly.

Essentially the transmitter consists of two little vibrating mirrors, E and F, one placed above the other. The lower mirror, E, of very narrow width, oscillates vertically at a frequency of 500 cycles per second. The upper mirror, which is somewhat larger, oscillates horizontally at about 10 vibrations per second.

A beam of light from the arc lamp, A, passes through the system of lenses, B, the aperture in the diaphragm, C, the lens, D, and impinges upon the lower vibrating mirror, E. From this mirror the beam, oscillating vertically 500 times per second, is reflected on to the top mirror, F, which is vibrating horizontally 10 times per second. The resultant action is that, as already explained in connection with other systems, we get a beam of light oscillating up and down and to and fro in a zig-zag path.

In the path of this beam there is placed the object the shadow outline of which it is desired to transmit, or a lantern slide, G. Passing through the slide the beam of light, made strong or weak in accordance with the transparent or opaque portions of the slide, is directed through the lens, H, which concentrates the fluctuating light beam on to the photoelectric cell, K.

SOME EARLY TELEVISION EXPERIMENTS 43

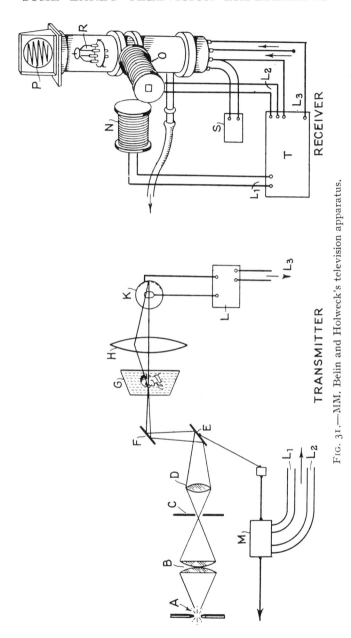

FIG. 31.—MM. Belin and Holweck's television apparatus.

The output of this cell is amplified and sent to the receiver, where it is caused to control the intensity of a cathode ray. The cathode ray oscillograph which is used in this apparatus is the work of M. Holweck.

This oscillograph, the external details of which are shown in Fig. 31, is very similar in arrangement and operation to the cathode ray tube already described in connection with Rosing's apparatus, although the details of the construction of the Holweck instrument are somewhat different. These details are shown in Fig. 32.

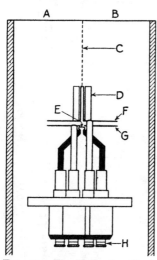

FIG. 32.—The Holweck cathode ray oscillograph.

Whereas the Braun tube used by Rosing was of the "cold cathode" variety requiring a potential of between 50,000 and 100,000 volts to operate it, Holweck's tube is of the "hot cathode" class, wherein the electron stream is provided by a glowing filament, as in a thermionic valve. This class of tube requires a very much lower voltage to operate it.

Referring to Fig. 32, above the filament, E, is placed a grid, G, which is a circular plate with a hole in the centre. Above this is a disc of similar form which acts like the anode or plate of a 3-electrode valve. This plate also has a hole in the centre, and over the hole is fitted a little copper tube. This arrangement concentrates the electron stream from the filament into a thin pencil-like discharge.

The filament is brought to the correct electron emission temperature by means of a 2-volt battery. The varying input current to the receiver (i.e. the amplified output of the photoelectric cell of the transmitter) is applied between the grid and the filament. The plate voltage required is 1500 volts, which is supplied by a special battery.

When this instrument is in operation, there is produced between the filament and the plate a stream of electrons which is "canalised" in the vertical tube, which is surrounded by a little coil, D, the current flowing through which produces the

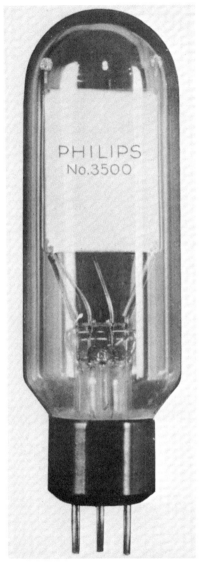

A typical flat plate type neon tube designed for television reception.

[*To face page* 44.

SOME EARLY TELEVISION EXPERIMENTS

necessary concentration of the stream by magnetic stress. The resultant fine ray strikes the fluorescent screen, AB, placed in the upper part of the oscillograph. The entire apparatus, shown in Fig. 32, is kept in a high vacuum by means of a Holweck molecular pump.

The action of the incoming current from the transmitter is to cause a disturbance in the normal emission of electrons, a disturbance which corresponds exactly with the variations of the modulated current at the transmitter. This disturbance is similar to that which takes place within a 3-electrode valve when a varying voltage is impressed upon the control grid. A positive charge on the grid attracts some of the electrons from the filament, thus reducing the number of them which reach the plate. The grid of the cathode oscillograph acts in the same way, increasing or diminishing the total number of electrons which reach the fluorescent screen, in accordance with the current variations received from the transmitter.

There remains the question of causing the electron stream to move to and fro in a zig-zag path across the screen. This is accomplished, in the Holweck tube, by the magnetic action of the two electromagnets, N and O, Fig. 31. One of these magnets is supplied from the same alternating current source which supplies the 500-cycle mirror at the transmitter, while the other is supplied from the same source as the 10-cycle mirror.

It will be seen that each of the magnetic fields has the effect of displacing the cathode ray in exactly the same way as the visible light beam is displaced by the oscillating mirrors at the transmitter. As this displacement is effected by means of the same current which drives the mirrors, it follows that synchronism will be obtained between the transmitter and receiver; but, of course, it is necessary to send the two alternating currents (500 cycles and 10 cycles) to the receiver as well as the picture impulses. Three separate channels are therefore necessary, although the two synchronising frequencies could be sent over one channel and filtered out at the receiver.

In the Holweck oscillograph the cathode ray, instead of being allowed to strike the screen directly, is first bent round at right angles by means of a prism. This is done for convenience, so that the screen, instead of being in a horizontal position on top of the instrument, is fitted in a vertical position which makes it easier to observe. The arrangement is seen in Fig. 31 at P, the zig-zag lines indicating the track of the cathode ray as it moves to and fro across the screen to construct the image.

CHAPTER V.

THE JENKINS SYSTEM.

Jenkins Prismatic Discs. Glow Lamps. The Moore Tube. Animated Shadowgraph Transmitter. Lens Disc. "Radiomovie" Receiver. Jenkins Drum Scanner. 4-Electrode Neon Tube. Advantages of Drum Scanning.

IN America the pioneer experimenter in the television field is C. Francis Jenkins, of Washington, D.C. Like Belin, Jenkins is well known in connection with phototelegraphy. He is also the inventor of the prototype of the modern cinematograph projector.

The Jenkins apparatus is distinguished by its originality, especially as regards the scanning mechanisms employed. In the systems so far described, light beams have been bent and moved about by means of mirrors. Jenkins decided to make use of prisms which, as explained in Chapter II., are also capable of performing the same service, and his solution of the problem, known as the Jenkins Prismatic Disc, represents an entirely new contribution to optical science.

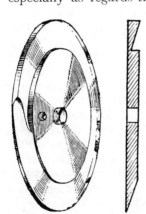

FIG. 33.—The Jenkins prismatic disc.

One of these prismatic discs is illustrated in Fig. 33. It consists essentially of a disc of thick glass, the outer edge of which has been ground into the shape of a prism, the section of which varies gradually and continuously round the circumference, so that at one point the base of the prism is outward, while diametrically opposite this point the base is inward, as shown in the cross-sectional drawing in Fig. 33.

If a beam of light is directed through the edge of such a

THE JENKINS SYSTEM

disc, it will be refracted or bent to an extent which depends upon the angle of the prismatic section at that point. For the sake of an example let us say that the beam is directed upwards at an angle of 45 degrees. If, now, the disc be slowly revolved, the beam will begin to move. If we are turning the disc in the right direction, the beam will move slowly downwards in a vertical plane until, when half a revolution of the disc has been made, the beam will be found to be pointing downwards at an angle of 45 degrees.

By superimposing a second disc over the first (as shown in Fig. 34), so that their overlapping edges revolve in directions at right angles to each other, a lateral movement can be given to the beam as well as a vertical movement. In this manner Jenkins obtained the two necessary oscillatory actions. It is essential, of course, for one of the discs to revolve comparatively slowly, to space the image lines traced over the light-sensitive cell, or on the screen, while the other disc must revolve much more rapidly in order to trace the image lines.

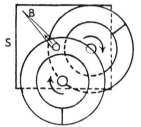

FIG. 34.—The arrangement of the Jenkins prismatic scanning discs.

Referring to Fig. 34, the lower disc, revolving slowly, imparts vertical movement to the beam of light, B, while the upper right-hand disc, revolving rapidly, imparts horizontal movement to the beam. The combined action of the two discs therefore causes the beam to sweep to and fro across the screen, S, covering it with light.

At the transmitting end Jenkins uses a magic lantern projector in which is inserted a lantern slide of the scene to be transmitted. The beam of light issuing from the lens of the lantern then passes through the prismatic sections of the revolving discs, which sweep images of the slide across the photoelectric cell.

At the receiver, the incoming impulses from the photoelectric cell are caused to control the degree of brilliancy of a special neon glow lamp. A beam of light from this lamp is then focused through a system of revolving prismatic discs similar to that in use at the transmitter, and these discs spread the light fluctuations over a screen, thus building up an image of the scene being transmitted.

Jenkins gave a public demonstration of this apparatus in Washington in June, 1925, transmitting over a distance of several miles, by wireless, the shadowgraph image of a slowly revolving windmill, which was substituted for the lantern slide.

Neon glow lamps, referred to above, are now in common use for advertising display purposes, where they take the form of long glass tubes which are bent into various designs, letters, etc., and glow at night with a brilliant orange or red glow. As various forms of glow lamps are now in general use for television purposes as well, it is perhaps advisable at this point to consider the subject briefly.

A stream of electrons, i.e. an electric current, will flow through a good conducting wire with ease ; the higher the resistance of the conductor the more difficult it is to force a current through it. Air is a high resistance conductor which requires something like 20,000 volts to force a current across an air gap one inch long. When a current does flow, it does so in the form of a spark. On a stupendous scale such a current flow, or spark, is called lightning.

The reason why the current flow, or discharge, is visible is because electrons forced from one of the solid electrodes collide with air atoms, and the impact knocks one or more planetary electrons out of their orbits. An atom minus one or more planetary electrons is in an unstable condition, and will sooner or later acquire fresh electrons to replace those which it has lost. It is the recombination process which produces the light which makes the discharge visible. Heat is given off at the same time, and this causes the air in the immediate vicinity of the discharge to expand. After the discharge has ceased, a partial vacuum is left as a result of the heat and consequent expansion. Since nature abhors a vacuum, air rushes in from all sides, meeting at the centre of greatest vacuum with such force that, in the case of a spark, a snapping noise is produced ; in the case of lightning, thunder is the result of the enormous disturbance.

If the air gap is sealed into a glass tube which is then slowly exhausted by means of an air pump, the discharge passes the more easily as the air pressure is reduced. In effect, the number of atoms of air present in the tube is being reduced, so that electrons forced from one of the electrodes has to travel farther before colliding with an atom, and thus acquires sufficient velocity to break it up under the influence of a lower propelling voltage.

[*Keystone Photo.*

J. L. Baird and some of his earliest apparatus, used in experiments at Hastings in 1924.

[*To face page* 48.

THE JENKINS SYSTEM

As the air pressure within the tube is reduced, so the spark fattens out and becomes quiet, until the discharge fills the entire tube with a glow known as the positive column. At this stage the primary electron stream breaks up atoms, and positive and negative parts of these join the stream and, in turn, break up other atoms. A condition of ionisation exists, ions being present everywhere in the tube.

If the pressure is still further reduced, the glow detaches itself from the negative electrode, leaving what is known as the Crookes dark space, and the entire surface of the electrode itself is covered with a glow which is called the negative glow. Further evacuation leads to other phenomena, and eventually to X-rays, but it is not necessary here for us to pursue the discharge through all its stages.

In passing, it is of interest to note the similarity between the above explanation and that given in connection with gas-filled photoelectric cells in Chapter III. It will now be understood why the voltage impressed upon the anode of a gas-filled photoelectric cell must not be too high. If it is raised beyond a certain critical value, the cell will "flash over," i.e. glow in the manner described above, due to ionisation of the gas atoms present in the cell.

Neon glow lamps are designed to make use of the phenomena described above. Two electrodes are sealed into a glass tube containing the gas neon at a suitable pressure. For advertising signs this pressure is adjusted so that the discharge is in the form of a positive column filling the whole tube. For television purposes two types are used, depending upon requirements. In one type the positive column is employed; in the other the gas pressure is adjusted so as to make use of the negative or cathode glow, i.e. the glow appears on the negative electrode only, the rest of the tube being in darkness.

In perusing this book, the reader will at some time or other commence to wonder why ordinary incandescent filament or arc lamps are not used as light sources in television receivers. The reason is not far to seek. Like selenium, such light sources will not respond rapidly enough. Some form of light is necessary, the brilliance of which will change, not only instantly, but also in exact proportion to the intensity of the incoming electrical impulses from the transmitter.

The glow discharge lamp is the only light source at present known which will conform to these requirements. It will respond to current impulses occurring at frequencies up to one million cycles per second, because its action is entirely electronic. Similarly, as the current intensity changes the number of ions produced also changes, the number of electrons falling back into atoms changes, and hence the intensity of the light produced changes in exact accordance with the current intensity.

Glow lamps can be made with gases other than neon. Both mercury vapour and helium are in common use, the former producing an intense greenish blue light and the latter a blue light. Neon tubes are used for television because, in spite of the low actinic value of the red light produced, they give the maximum light for a minimum current input. The manufacturing problems of neon tubes are also simpler, which makes for cheapness.

In operation, the neon tube is connected to the output stage of the wireless receiver which is receiving the television signals, but it is first "biassed" with a voltage just sufficient to cause it to glow. The actual biassing voltage required depends upon the characteristics of the tube in use, but for the tubes generally used in this country the voltage is of the order of 160 volts. The just glowing tube is then influenced by the incoming signals, which increase or decrease the brilliance of the glow in accordance with their intensity.

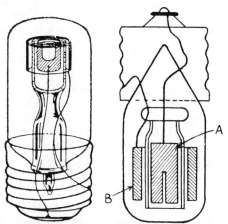

FIG. 35.—Two views of the crater type neon tube developed for Jenkins by D. MacFarlan Moore.

Two views of the type of neon tube employed by Jenkins in conjunction with his prismatic disc apparatus are given in Fig. 35. This tube, known as the crater type, was developed for Jenkins by D. MacFarlan Moore. The outer electrode, shown at B in the sectional sketch, is the negative electrode, while the inner electrode, A,

THE JENKINS SYSTEM

is positive. They are separated by a tube of glass which is open at one end. The discharge, of the positive column variety, is concentrated in the centre, or crater of the positive electrode, and is of great intensity. This form of design is adopted in all cases where it is desired to produce a concentrated high intensity spot discharge for focusing through optical systems.

In his earlier experiments in connection with photo-telegraphy, Jenkins secured synchronism between his transmitting and receiving mechanisms by means of an electrically driven tuning-fork arrangement. As will be explained later on, such an arrangement is not sufficiently accurate for television purposes, and for television work Jenkins now makes use of synchronous motors to drive both transmitter and receiver mechanisms. These motors are supplied with current from the 110 volt, 60 cycle A.C. mains, which are standard almost everywhere in America.

Details of such motors will be given in a later chapter. For the present it is sufficient to say that two or more such motors, running off the same alternating current supply, *must* run at the same speed. Since large American manufacturers of alternating current generators claim that in all their large installations they can guarantee that the difference in frequency between any two plants is less than one-half per cent., this, according to Jenkins, guarantees the accuracy of operation of his television machines in all localities where the standard 60 cycle alternating current is available; for the only factor which will cause a synchronous motor to vary in speed is a variation in the frequency of the supply.

At this low frequency, however, the phenomenon known as phase swinging may prove detrimental to reproduction; for although the mean speed is constant, an A.C. synchronous motor, if fed from a low frequency supply, may vary its instantaneous speed quite considerably, and any such variation produces a very distressing swinging or "hunting" effect in the received image which is very trying to the eyes of the observer. Momentary speed variations may also be caused by the sudden application of a heavy load on the mains locally.

In a later type of transmitting apparatus Jenkins uses a lens disc carrying forty-eight lenses, and the purpose of the apparatus is avowedly to transmit and receive, not television, but special cinematograph films. There is no detail in the films, only a plain black-and-white silhouette of simple scenes such as a little girl bouncing a ball.

The films are reeled off through the transmitter at the rate of fifteen "frames" or pictures per second, and received at the receiver at the same rate. The general layout of the animated shadowgraph transmitter is shown in Fig. 36. The film reels are mounted on a simple framework, one above the other, in such a manner that the film is pulled downwards by a set of sprockets which are driven by an electric motor. One end of the shaft which drives the sprockets is fitted with a gear pinion which meshes with a smaller one on the shaft of the motor, which is a synchronous A.C. motor running at 1800 R.P.M. Because of the 2 to 1 reduction gearing the pictures are pulled past the sprockets, or any fixed point next to the film, at the rate of 900 per minute, or 15 per second.

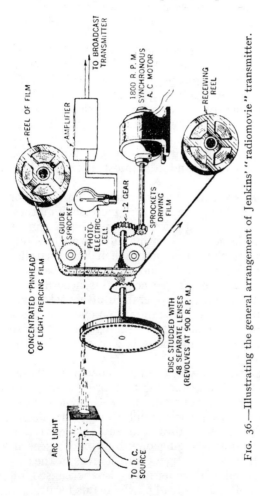

Fig. 36.—Illustrating the general arrangement of Jenkins' "radiomovie" transmitter.

At the other end of the shaft which drives the sprockets is a heavy metal disc, about 15 inches in diameter and about 1 inch thick. The edge of this disc is studded with forty-eight

THE JENKINS SYSTEM

separate little lenses, each of which has an "optical speed" of $f/3\cdot5$. These lenses are designed to concentrate the light from a powerful arc lamp into an intensely brilliant pin-head beam, which is caused to pierce the film as the latter travels down past the back of the disc.

Immediately behind the film is mounted a photoelectric cell, which is so placed that the beam of light, after passing through the film, impinges upon it. The output of the cell is amplified in the usual manner and passed on to the wireless transmitter.

A close study of Fig. 36 will make the operation of the apparatus clear. The lens disc revolves at 900 R.P.M., or fifteen times per second. The separation between the centres of the lenses is just equal to the width of the film. The latter moves steadily downwards at the rate of fifteen pictures per second. Its action is not jerky, as in a cinematograph projector, but smooth.

The arc lamp on the left projects a powerful converging beam of light through one of the lenses of the disc, which lens further converges the beam to a point, which scans or travels across the moving film from one side to the other, due to the rotary motion of the disc. When the next lens picks up the beam from the arc lamp, the film has moved downwards slightly, so that the second beam travels across the film on a parallel but slightly higher path.

Succeeding lenses of the disc trace further parallel paths across the film, until each picture has been explored by the forty-eight lenses. As soon as the beam of light from one lens runs off the film, the beam from the succeeding lens starts to pierce the film on the other side. This movement is continuous during the operation of the mechanism. Thus forty-eight separate beams of light travel across each individual picture in one-fifteenth of a second. At the start of each fifteenth of a second period a fresh picture slides into position, and another series of forty-eight light beams starts to pierce it.

While this movement is taking place, the light beams, after shining through the film, fall on the photoelectric cell with degrees of intensity which depend upon the density or opacity of the parts of the film through which they shine, and the intensity of the output of the photoelectric cell varies accordingly.

The receiver for working in conjunction with this transmitter has many meritorious points of difference from any other receivers described in this book. It is illustrated in Fig. 37, and consists of six essential parts. The heaviest unit is a 3600 R.P.M. synchronous A.C. motor, to the shaft of which is attached a hollow metal drum 7 inches in diameter and 3 inches in length. The centre of this drum is a hollow spindle with a thin wall, having an inside diameter of $1\frac{1}{2}$ inches.

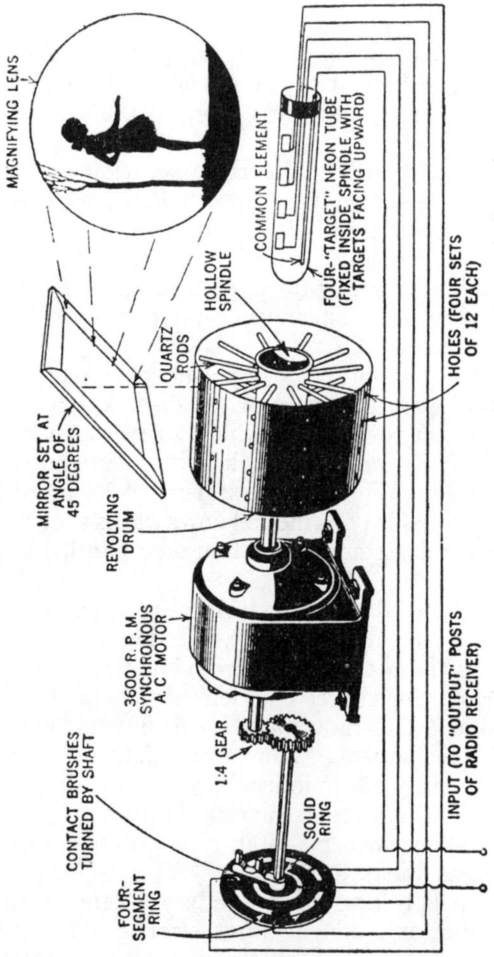

In corresponding places on the drum and the spindle are four spiral rows of twelve holes, each $1/32$ inch in diameter. The spirals are spaced half an inch apart and the holes in each drum spiral are spaced 2 inches apart circumferentially. A short piece of quartz rod between the drum and spindle connects each pair of corresponding holes. The purpose of the forty-eight little quartz rods is to conduct light from the inner spindle to the holes in the outer drum without loss, quartz having the peculiar property that light flows through it like water through a pipe. The use of such rods thus avoids loss of light due to the inverse square law referred to in Chapter II.

FIG. 37.—General arrangement of the Jenkins' "radiomovie" receiver.

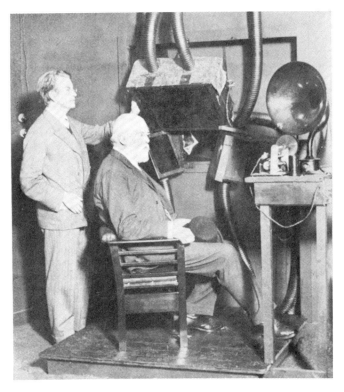

[Courtesy: Baird Television Ltd.

Sir Oliver Lodge examining Baird's noctovision transmitter at the British Association Meeting at Leeds in 1927.

[To face page 54.

THE JENKINS SYSTEM

Fixed inside the hollow spindle, with the little flat plates facing outwards and upwards, is a special neon tube having four discharge electrodes. This tube has an outside diameter of $1\frac{1}{8}$ inch, and the electrodes measure $3/_{16}$ by $\frac{1}{4}$ inch. A straight wire running near the four electrodes acts as the common positive electrode, for this tube is designed to make use of the negative glow on the surface of each negative electrode. In Fig. 37 the tube is shown withdrawn from the spindle, in order to illustrate it; in actual use it is mounted in a clamp fixed on the motor platform, and fits snugly inside the revolving spindle, without touching it, in such a manner that each negative electrode is opposite one of the spiral rows of holes.

The other end of the motor shaft is fitted with a 4 to 1 reduction gearing which drives a revolving switch. The revolving element is simply a pair of contact brushes connected together. One brush effects continuous electrical connection with a solid brass ring imbedded in an insulating disc, while the other makes a wiping contact over the four sections of a split ring. The four segments are connected to the four discharge electrodes of the neon tube, while the solid ring is connected to one of the output terminals of the wireless receiver. The common electrode of the neon tube is connected to the other output terminal of the radio receiver.

All the receiving apparatus described so far is contained in a wooden box measuring about 2 feet long and a foot square at the ends. Directly above the top of the revolving drum is a square opening in the top of the cabinet, and over this opening an ordinary mirror is mounted at an angle of 45 degrees to the top. About a foot in front of the mirror, and standing upright, is a magnifying lens about 10 inches in diameter.

Continuing our explanation of the action of the receiver, the modulated picture signals from the distant transmitter are picked up by an ordinary wireless receiver, amplified, and fed to the "radiomovie" receiver, as Jenkins has named his instrument. Assuming that the contact brushes have just made contact with the upper right-hand ring, as shown in Fig. 37, and that one of the quartz rods in the first, or outermost spiral of holes, is standing straight up, this condition corresponds with the start of a picture in the transmitter, when the light spot is just commencing to sweep across the film.

As the contact brushes have just closed the circuit to the neon tube electrode at the extreme right, this electrode lights up immediately as the biassing voltage reaches it, and thereafter fluctuates in brilliancy exactly in accordance with the modulation of the incoming picture impulses. The fluctuations of light are carried up the quartz rods and projected through the holes in the outer drum on to the mirror. The light thus reflected from the mirror follows the shading of the images on the original film, so that as the drum revolves a picture is built up in the mirror. This picture may then be observed through the magnifying glass.

A complete picture of forty-eight lines (corresponding to the rate of transmission) is built up on the mirror with every four revolutions of the drum. At the beginning of the second revolution, the contact brushes turn to the next segment of the switching ring (because of the reduction gearing) and the signal impulses are transferred to the second electrode of the neon tube. The third and fourth quarters of the picture are similarly built up from the third and fourth neon electrodes, and the cycle then commences all over again. During one second the drum revolves sixty times. Since four revolutions create one picture, sixty revolutions create fifteen pictures, which gives the speed of fifteen pictures per second mentioned when the action of the transmitter was being described.

As in the case of the prismatic disc apparatus, Jenkins relies for synchronism upon the use of synchronous motors running off the standard A.C. mains. Witnesses describe the pictures, as viewed at a distance of a few feet from the magnifying lens, as being clean-cut black silhouettes against the characteristic reddish background provided by the neon tube, and all movements in the film scene are faithfully and smoothly portrayed. Unmagnified, the pictures are 2 inches square; viewed through the magnifying glass they appear to be about 6 inches square.

This apparatus is interesting and distinctly original, but it should be clearly understood that it does not produce television, which is the instantaneous vision of distant scenes or events while they are actually in progress. In other words, the transmitting apparatus must operate from life—not from a prepared record such as a film or lantern slide of an event long past. In this Jenkins apparatus, using silhouette films, there are no refinements of shading and detail to be handled. All the photoelectric cell has to do is to distinguish between a strong direct light and no light at all. As has already been explained, the

THE JENKINS SYSTEM

problem of designing apparatus suitable for the accomplishment of true television is much more difficult.

The special advantages claimed by Jenkins for his drum scanner and associated multi-target neon tube are, firstly, that a drum scanner is much more compact and requires less power to drive it than the disc scanner which will be dealt with in the next chapter; his 7-inch drum, giving a picture 2 inches square before magnification, corresponds to a 48-hole disc measuring 36 inches in diameter. Secondly, Jenkins claims that far less power is required to supply his 4-electrode neon tube than would be required to supply the equivalent flat plate neon tube which is used with a disc scanner. The former requires only 3 to 5 milliamperes of current, whereas the latter, with a plate measuring $2\frac{1}{2}$ inches square, requires about 60 milliamps. This difference is accounted for by the fact that, in the Jenkins tube, only a tiny electrode area needs to be lighted at any given instant, whereas in the case of the flat plate neon the entire plate must be kept constantly lighted, although only one tiny elemental area (about $1/_{32}$ of an inch in diameter) is in service at any given instant. The heavier current demands of the flat plate neon mean, of course, considerably increased complication and first cost in the L.F. amplifier of the wireless receiver, and higher running costs.

CHAPTER VI.

THE BAIRD SYSTEM.

Early Efforts. Nipkow Disc Scanner. Spot Light Transmitter. Flat Plate Neons. Lens Disc Scanner. Flood Light Transmitter. Optical Lever. Relative Motion. Multiple Channels.

WE must now turn to a consideration of the work done in England by John Logie Baird, to whom belongs the credit of being the first to demonstrate publicly true television. Half a century ago the inventive genius of Alexander Graham Bell, a Scotsman, gave to the world the telephone, and now Baird, another Scotsman, in successfully demonstrating television, has given us an electric eye to add to the electric ear provided by Bell.

It was in 1923 when Baird, compelled by ill-health to abandon an active business career, commenced to devote himself exclusively to a study of the problem of television. At first glance the problem seemed simple of solution. Briefly reviewed, the requirements, as we have already seen, may be stated as follows :—

(1) Means of scanning an image, so as to subdivide it into tiny sections, or elements.

(2) Means of transforming the resulting picture elements, or light impulses, into electrical impulses which can be transmitted to the distant receiver, either by wire or by wireless.

(3) Means of reconverting electrical impulses into light impulses, and by means similar to (1), causing them to cover, or illumine a screen, thus reproducing the image at the transmitter.

(4) Means of synchronising the transmitter and receiver mechanisms.

On paper the problem seemed to Baird to be simple enough. Many optical methods were already known which would fulfil requirement (1). The selenium cell and the photoelectric cell were in existence and seemed to cover requirement (2) ; and for (3) there was the neon tube. Methods of synchronism had

THE BAIRD SYSTEM

already been developed to a high degree of accuracy in connection with other electrical processes, notably as an aid to highspeed multiplex telegraphy, so that point (4) seemed to be covered.

It all seemed very simple, so simple, in fact, that the wonder was that television had not already been accomplished. It did not take Baird very long to discover wherein lay the difficulties. The lag of selenium presented an obstacle. Photoelectric cells were not nearly so perfect then as they are to-day, and it was not possible to obtain the necessary degree of amplification by means of thermionic valve amplifiers. Another difficulty was that the available methods of synchronism, although satisfactory for the purposes for which they were developed, were quite unsuitable for television.

After about six months' work, however, Baird achieved a measure of success, and was able to transmit shadowgraphs. Working alone and with apparatus of the crudest description, he persevered until, in April, 1925, he had the satisfaction of giving the first public demonstration of the electrical transmission of silhouettes between two separate machines at Selfridge's store in Oxford Street, London. As stated, only silhouettes were transmitted on this occasion, but they were transmitted by light reflected from the original objects, and not by directly transmitted light as had been done previously.

After several more months of hard work, Baird was able, on January 27th, 1926, to give to over forty members of the Royal Institution the first demonstration of true television ever witnessed. The images of moving human faces were transmitted between two rooms in Frith Street, Soho, not as outlines or silhouettes, but complete with tone gradations of light and shade and detail. Although the received images flickered a great deal, and suffered from other defects, the individual being " televised " could just be recognised.

The original transmitting mechanism is now on exhibition in the South Kensington Science Museum. It is a weird-looking collection of old bicycle sprockets, cardboard discs, and bull's-eye lenses, all tied together with sealing wax, wire, and string.

In previous chapters we have examined several methods which have been devised from time to time to scan an image. The simplest scanning mechanism ever produced was invented as long ago as 1884 by a German scientist named Nipkow, who at the same time laid down the requirements for the achievement of television, but was not able to achieve success himself because of the lack of suitable apparatus.

The Nipkow disc consists simply of a disc of thin sheet-metal, around the periphery of which is drilled a single spiral of small holes, each hole being displaced radially one hole diameter nearer the centre of the disc than its predecessor in the spiral. When vertical scanning is employed, the radial distance between the first and last holes of the spiral governs the width of the area which can be scanned, while the circumferential distance between successive holes governs the height of the scanned area.

Referring to Fig. 38, where the Nipkow disc is shown at B,

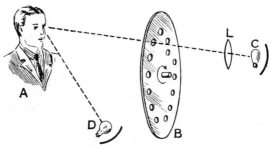

Fig. 38.—Illustrating the Nipkow disc and its operation in the light-spot system of scanning.

a beam of light from the lamp, C, is focused on to the edge of the disc in such a way that it can only pass through one of the holes at a time. If, after passing through one of the holes the light beam is allowed to fall on a blank white sheet, or screen, the observer will see just a single spot of light. If the disc is now rotated very slowly about its axis, the light-spot will be seen to move slowly in a vertical direction. Just as the spot leaves the edge of the screen (either at the top or bottom, depending upon the direction of rotation of the disc), a second spot will appear at the opposite edge and, owing to the staggered, or spiral arrangement of the holes in the disc, a little to one side of the path of the first spot. As this second spot leaves the screen a third spot will appear at the opposite edge, again a little to one side, and so on.

When one full revolution of the disc has been accomplished,

[*Courtesy: Baird Television Ltd.*

J. L. Baird with his improved "noctovisor," arranged as an aid to navigation during foggy weather.

it will be found that every part of the screen has been covered, in turn, by the light-spot.

If, now, the disc is rotated a little faster by means, say, of an electric motor, it will be observed that the light-spot begins to travel so rapidly that it can no longer be followed on the screen as a spot ; it becomes a white streak which moves slowly but steadily from one side of the screen to the other. As the streak passes off one side of the screen, it instantly reappears again on the other side, moving across it. It is because of persistence of vision that the spot tends to appear as a streak when it is moved in excess of a certain speed.

If the disc is speeded up a little more, a speed will be reached (about eight revolutions per second) when even the travelling streak cannot be followed, and at this critical speed the *entire screen* appears to be steadily and continuously illuminated all over by parallel vertical streaks of light which give to the eye a sensation similar to the " pouring rain " effect of the early cinematograph.

If the revolving disc is now stopped dead, the observer immediately realises what tricks persistence of vision has played on him, for instead of a steadily illuminated screen he finds himself once more looking at a tiny stationary spot of light.

One factor which will immediately be apparent to him is that the apparent intensity of the illumination of the screen when the disc is revolving is considerably less than the intensity of the light-spot when it is stationary. This is not due to any diminution in the intensity of the spot when it is moved ; it is due simply to the fact that, while visual impressions do not fade from the retina of the eye instantly, they *do* fade in the course of time and (at the speed given above) one-eighth of a second elapses from the time the observer sees, say, the top left-hand corner of the screen illuminated till he receives a visual impression of the light-spot at the lower right-hand corner. This partly explains a phenomenon of television reception, i.e. the great disparity between the apparent brilliance of the received image and the intensity of the neon lamp which is used as the light source.

To return to our explanation of the light-spot transmitter, if, instead of the screen we have used for purposes of explanation, we substitute the head of a human being, it will be obvious to those who have followed what has been written above that the light-spot will, in the course of one full revolution of the disc, shine upon every part of the features of the person being televised. At any given instant light is reflected back from the particular part of the face upon which the light-spot is resting at that instant, and this minute point, or flash of reflected light, is arranged to fall on a photoelectric cell (or group of cells) placed as shown at D in Fig. 38.

The intensity of the reflected light at any given instant depends entirely upon the lines, contours, and colour (or degree of light and shade) of the particular part of the face upon which the light spot is falling at that instant ; and upon the intensity of the reflected light depends the magnitude of the currents delivered by the photoelectric cell, or cells.

Such is the manner in which, by this system of scanning, an image of the scene to be transmitted is transformed into electrical impulses which vary in strength from instant to instant in strict accordance with the light and shade and outlines of the scene.

At the receiving station the process is reversed. The incoming electrical impulses, after amplification, are fed to a neon lamp, placed like C in Fig. 38, but close up to the back of the disc. The arrangement is such that the observer, looking at the front of the disc, can only see the neon lamp through one hole in the disc at a time. He therefore sees, when the disc is stationary, a single tiny but intensely brilliant spot. When the receiving disc is revolved at precisely the same speed as the transmitting disc, and exactly in phase with it, the observer watching the glowing neon lamp through the holes of the disc sees an exact representation of the scene which is being televised. The type of neon tube used, known as the flat plate type, makes use of the negative glow, which appears to cover the surface of the cathode. This cathode is simply a flat plate of metal the area of which is slightly larger than the scanning area (height to width ratio of the arrangement of the holes) of the disc. A typical flat plate neon tube will be found in the photograph facing page 44.

It only remains to be added that the greater the number of holes there are in the Nipkow scanning disc, and the smaller they are, the larger will be the number of strips into which the image is divided, and the strips will be narrower, thus giving to the received picture a larger amount of detail.

All optical systems can be reversed, and a reversal of this

THE BAIRD SYSTEM

system was used by Baird in his first experiments, using flood lighting instead of spot lighting. In this system the Nipkow disc with its single spiral of tiny holes is replaced by a much larger diameter disc around the periphery of which is mounted, again in the form of a single spiral, a series of lenses. Such a disc can be seen in the photograph facing page 36.

The arrangement of a transmitter operating along these lines is shown in Fig. 39. The scene to be televised is illuminated by flood lighting, as in a photographic studio, by the lamps, L, or by ordinary daylight. The lenses in the revolving disc, D, then cast images of the scene in rapid succession on to the photo-electric cell, C. Prior to falling on the cell the images pass through shutters (not shown in Fig. 39), whose purpose is to

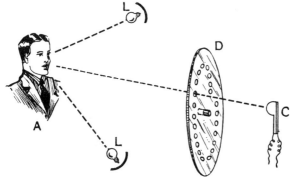

FIG. 39.—Illustrating the spiral lens disc and its operation in the flood-light system of scanning.

ensure that successive parts of the image fall on the cell in regular order.

When a lens disc is employed at the receiver, the arrangement is exactly similar to that adopted in the case of a Nipkow disc, the only difference being that a larger neon lamp is required, owing to the fact that, in order to accommodate the lenses, a much larger disc has to be used, and the radial distance between the outermost and innermost lenses in the spiral is greater than in the case of a Nipkow disc, thus giving a larger image. A larger neon tube requires more power to supply it, which means a larger and more expensive amplifier. Since, at the receiver, equally good results can be secured by means of a Nipkow disc and a smaller neon lamp, there is no advantage in using a lens disc for reception purposes, for the received image can be enlarged by means of a single magnifying lens placed in front of the disc. For transmission purposes, however, there is an advantage in using a lens disc when it is desired to televise extensive scenes, or scenes illuminated by daylight.

Another interesting scanning device, due to Baird, is a somewhat unusual honeycomb structure made up of a bank of small tubes arranged in parallel rows. The action of this bank of tubes is rather peculiar.

If held, for example, in front of an electric light bulb an image of the filament will appear on a ground glass screen held close against the back of the bank of tubes. A surprising feature of the image is that it is constant in size, independently of the distance between the bank of tubes and the lamp. In the case of all other image-casting devices, the size of the image is dependent upon the distance away of the object from the image-casting device.

Whether this particular apparatus will find a permanent use in television is perhaps doubtful, but it is described here as an interesting scientific novelty which has been tried experimentally as a substitute for a revolving lens disc.

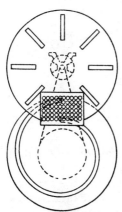

FIG. 40.—Baird's cellular structure scanning device.

Referring to Fig. 40, two interrupter discs are arranged to operate in conjunction with the cellular structure, so that its surface is scanned in such a way that light is admitted to the photoelectric cell through one tube at a time. As already explained, disc receivers give an image which appears to be made up of thin closely fitting strips. By using the system of projection tubes mentioned above, the received image takes on the appearance of being made up of a large number of fine dots.

Still another highly interesting scanning device, also due to Baird, has been called by him an "optical lever." In order to obtain a finer grain in the received image, i.e. greater detail when using a lens disc or a Nipkow disc for scanning purposes, it is necessary to increase the number of lenses or holes in the disc. In order to retain the same size of image this entails increasing the diameter of the disc, and a point is soon reached where

THE BAIRD SYSTEM

the proportions of the disc make it not only extremely unwieldy, requiring an unduly large amount of power to rotate it, but, in the case of a lens disc, centrifugal force prevents it being rotated fast enough. The optical lever was designed to overcome all mechanical difficulties, and by means of it there is no limit to the speed at which light impulses may

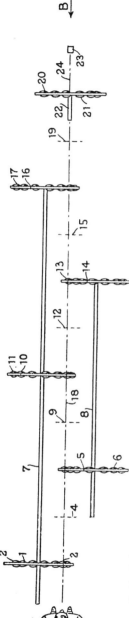

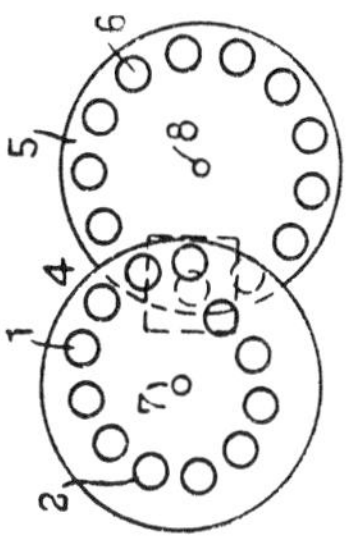

Fig. 41.

Fig. 42.

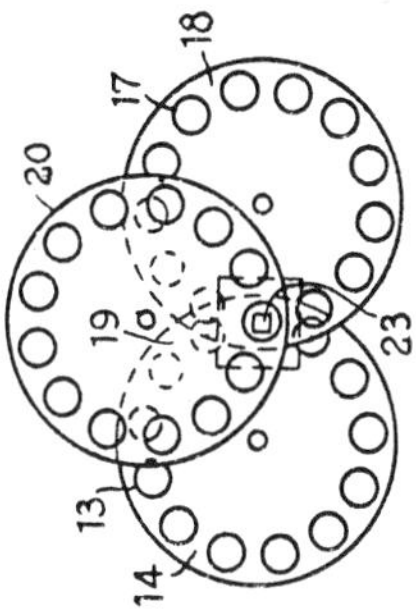

Fig. 43. — Baird's optical lever.

be caused to traverse the light-sensitive cell, or a bank of several cells.

To obtain a finer grain in the image it is necessary, as already explained, for it to pass very rapidly over the cell in finely divided elemental areas, and in order to increase the speed of traversal without increasing the speed of revolving discs, or adding lenses, more than one combination of discs is used.

Referring to the accompanying

diagrams of the optical lever (which are reproduced from the patent specification), Fig. 41 is a plan of the apparatus, Fig. 42 is a partial end view looking in the direction of the arrow A, and Fig. 43 is an end view looking in the direction of the arrow B. In the following description, ground glass screens will be mentioned, as they enable the action of the system to be followed more easily. Actually they are not essential to the performance of the apparatus, and are omitted in practice.

Referring to Fig. 41, the image-scanning mechanism at the transmitter consists of a revolving disc, 1, in which is mounted any desired number of lenses, 2. These lenses are mounted in a circle, not a spiral. In front of the disc is the object, 3, of which a reproduction is to be transmitted. The position of the first ground glass screen, which is placed at the focal point of the lenses, 2, is shown at 4. The line 18 indicates the optical axis of the apparatus, along which projection of the images takes place. When one of the lenses, 2, crosses this axis an image is projected upon the screen, 4.

Behind the screen, 4, is a second revolving disc, 5, exactly similar to the first disc, and carrying a similar number of lenses, 6. This second disc is so positioned that its lenses, by the rotation of the disc, are carried across the optical axis, 18. The two discs are mounted on separate parallel shafts, 7 and 8, which are rotated at any desired speeds. They may, if desired, be geared together.

The image projected on to the screen, 4, constitutes an object of which an image is projected by each of the lenses, 6, on to a screen, 9, placed behind the disc, 5; and the direction of rotation of the shafts, 7 and 8, is made such that the lenses, 2, cross the optical axis, 18, in the opposite direction to the lenses, 6. With the construction shown, this result is obtained by rotating the discs in the same direction.

The two shafts carry a further series of similar lens discs, and the lenses on each disc, as they cross the optical axis, reproject successively from the left towards the right (in Fig. 41) the image cast by the lenses on the preceding disc in the series, and as the lenses are moving successively in opposite directions, the speed of traversal of each successive image is increased. Any desired number of discs may be used on each shaft so as to provide any desired final speed of traversal of the image.

The most difficult point to understand in connection with this mechanism is just how it comes about that the speed of

[*Fox Photos.*

Jack Buchanan, the well-known actor (left), posing before Baird's daylight television transmitter.

[*To face page* 66.

THE BAIRD SYSTEM

traversal is increased. At first glance it would appear that no advantage would result from increasing the number of discs, and that the images cast by the lenses in successive discs would, in any case, just pile one on top of the other on the last screen of the series. Such, however, is not the case.

All motion is relative. In order to decide what is motion and what is not, and the speed at which an object moves, we must have some fixed point of reference. A person may walk along the corridor of a train at a speed of three miles an hour. Persons sitting in the compartments would say he was moving at the rate of three miles an hour, which would be correct from their point of view, i.e. using themselves as the point of reference.

But if the train itself were travelling at sixty miles an hour, and the person concerned happened to be walking from the back of the train towards the front, his total rate of progress, using the surface of the earth as the point of reference, would be sixty-three miles per hour.

When we consider that the earth itself is moving through space, revolving round the sun, and round its own axis, then we see that the rate of progress of the person walking along the train, and the speed of the train itself, if we take the sun as the point of reference, is something altogether different to what we imagined it to be.

Returning now to the optical lever, the lenses of the first revolving disc cast on the screen, 4, a series of images of the relatively fixed object, 3. The lenses of the second disc, which cross the optical axis in the opposite direction, pick up an image of screen, 4, and cast it on to screen, 9. From the point of view of the second disc, the image on screen, 4, is a relatively fixed object, a series of images of which it has to project on to screen, 9. But by virtue of the fact that the image on screen, 4, is not a fixed object but a series of rapidly recurring images ; and by virtue also of the fact that the lenses on the first and second discs are moving in opposite directions, the total number of images which will fall on screen, 9, in a given time will be twice the number falling on screen, 4. The second disc really handles images at the same speed as does the first disc, but whereas the latter is projecting images of a fixed object, the former is, as it were, projecting images of a number of rapidly moving objects. The same reasoning applies to each successive stage.

Thus the speed of traversal of the images is increased by what may be described as an optical lever effect ; hence the name which has been given to the device. At each successive screen the speed of traversal of the images is twice what it was on the preceding one, and the increase may be carried to any desired extent by adding more discs to the series, whilst maintaining the speed of the mechanism itself within permissible limits.

As already stated, screens are not actually necessary; they have been quoted in the description given above, and shown in Fig. 41, simply as an aid to the understanding of the operation of the mechanism. As they cross the optical axis the lenses on each disc will pick up the images cast in space by the lenses of the preceding disc.

With the arrangement shown, the final screen, 19, has cast upon it a series of images which are moved at a high speed upwards (or downwards, according to the direction of rotation of the discs), and if a photoelectric cell were substituted for this screen it could be moved slowly across the moving image to produce the desired exploration of the entire image. This relative lateral movement between the series of images and the cell is preferably effected by means of a separate disc, 20, which is mounted on a separate shaft, 22, situated above the plane of the shafts, 7 and 8, so that the disc, 20, overlaps the optical axis, 18, as shown in Fig. 43.

The disc, 20, carries a series of lenses, 21, which are moved across the optical axis in a direction at right angles to the direction of motion of the lenses, 14 and 17. The shaft, 22, may be rotated slowly compared with the shafts, 7 and 8, thus causing the images to move laterally across the photoelectric cell, so as to provide an adequately finely grained picture. The photoelectric cell, 23, is placed on the optical axis as shown.

The lateral movement of the series of images may be effected in any other desired manner. Thus, for example, the lenses on one of the discs could be arranged in a spiral, as shown on the disc, 1, in Fig. 42, instead of in a circle, so that each of the succession of images produced by the lenses of that disc is displaced slightly to one side of the previous image.

The principal disadvantage of this ingenious invention is that so much light is lost in passing through this extensive optical system that it is doubtful whether sufficient would be left at the end of it to give rise to an adequate response in the photoelectric cell. This objection places a limit to the number of discs which can be added to the series. In any case, the number of impulses to be handled by the photoelectric cell per second would be far

THE BAIRD SYSTEM

in excess of the ability of present-day transmission channels to accommodate. This subject will be dealt with more fully in a later chapter.

Baird has, however, described an arrangement whereby more than one photoelectric cell is employed, each cell being arranged to deal with its own band, or section of the picture, and control its own light source at the receiver, where each light source reproduces the section of the picture dealt with at the transmitter by the cell which controls it. The signals, in this case, are sent along separate lines, or on separate wavelengths, if wireless transmission is used. This idea has been put into effect recently by Sanabria, an American experimenter.

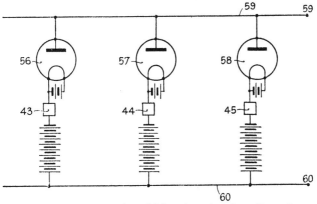

Fig. 44.—The arrangement of multiple photoelectric cells and associated 2-electrode valves suggested by Baird.

If the transmission medium will handle impulses at the required speed, however, there is no reason why the outputs of the photoelectric cells should not all be transmitted over the same circuit, or channel, the output impulses of each cell being timed electrically to follow those of the preceding cell in the series, all in the proper order. By the use of appropriate signal-accepting circuits at the receiver, the incoming impulses could be sorted out and applied to their proper light sources.

In cases where more than one photoelectric cell is employed, each cell should be provided with its own amplifier. Even after individual amplification, it is advisable, according to Baird, to place a 2-electrode valve in the output circuit of each amplifier to prevent feed-back. Alternatively a 2-electrode valve can be placed in the output circuit of each cell direct, whereupon the output circuits of the three 2-electrode valves can safely be connected in parallel to a single amplifier. This latter arrangement is shown in Fig. 44 (which is reproduced from the patent specification), where 43, 44, and 45 are the photoelectric cells ; 56, 57, and 58 are the 2-electrode valves ; and 59 and 60 are the output terminals which are connected to the input terminals of the amplifier.

CHAPTER VII.

THE BAIRD SYSTEM (*continued*).

Phonovision. Listening to a Face. Phonovisor *v.* Talkies. Noctovision. The Spectrum. Generating Infra-red Rays. Penetrative Qualities of Infra-red Rays. Noctovisor as aid to Navigation. Television by Daylight, in Colours, and in Stereoscopic Relief. Enlarging the Received Image.

SINCE 1927 Baird has from time to time staged spectacular demonstrations of either television itself or variants and sidelines of it, all of which are of such scientific interest that this chapter will be devoted to a consideration of these novelties, and the apparatus and methods by means of which they have been achieved.

One of Baird's first " by-products " was called by him " phonovision." This consists essentially of the recording of a television image on a gramophone disc, so that it can afterwards be played over and the recorded image made visible in a television receiver.

If, while a television transmitter is in operation, a pair of head-phones is plugged into the photoelectric cell amplifier the current impulses from the cell will be heard as a characteristic note. In other words, if a human being is seated before the transmitter, one finds oneself listening to a human face ! When the television image signals are broadcast by wireless, they can be heard on a loud speaker in exactly the same manner, and no doubt most readers of this book will have heard these signals during the daily broadcasts of television by the B.B.C. It is interesting to note that each face or scene produces its own characteristic note, and after a period of close study it is quite possible to distinguish between one face and another by the sound alone.

Instead of broadcasting the output of the transmitter, or sending it along a wire line, Baird conceived the idea of feeding the signals to the electrically-operated cutting stylus of a gramophone recording machine, and causing it to cut a record

72   FIRST PRINCIPLES OF TELEVISION

of the scene on the surface of a wax master disc mounted on a turntable driven through gearing by the same motor which drives the scanning disc. The layout of the arrangement is illustrated in Fig. 45. From the wax master permanent discs are made in the usual manner.

If a phonovision record is played over in an ordinary gramophone, it reproduces a sound exactly like that heard during the television broadcasts. If, on the other hand, an electromagnetic pick-up is used instead of a sound box, and the amplified output of the pick-up is fed to a television receiver, the recorded image will be seen. To do this the turntable is driven, through gearing, by the same motor which drives the scanning

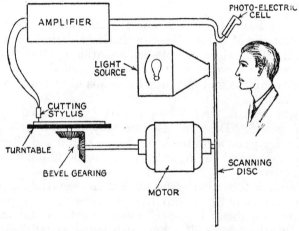

FIG. 45.—General arrangement of phonovision recording mechanism.

disc of the television receiver, or "televisor." The layout is very similar to that shown in Fig. 45, an electric pick-up being substituted for the cutting stylus, and a neon lamp for the photocell. The lamp is mounted behind the disc in place of the light source.

There is no difficulty about synchronism in phonovision, for both during recording and reproduction the scanning disc and turntable are driven by the same motor. It is only necessary, before commencing to reproduce, to see that the record is so positioned on the turntable that the pick-up needle rests on the correct starting-point. This starting-point on the record should coincide with a starting-point marked on the turntable,

[*Photopress Photo.*

Mrs. Howe (left, seated before transmitter), the first woman whose image was televised across the Atlantic, Feb. 9th, 1928. Seated behind Mrs. Howe is the author.

[*To face page* 72.

THE BAIRD SYSTEM

otherwise the position of the record will not be in phase with the position of the scanning disc.

The Phonovisor, although as yet in a very crude state of development, offers an alternative to the cinematograph film as a method of recording moving pictures. The voice of a sitter can be simultaneously recorded on a second record, synchronised with the phonovision record; or by means of a double track record both voice and picture might be recorded on the same disc.

In view of the present highly developed state of the talking film, however, it is unlikely that phonovision will have much influence. There is also this further difficulty. With television in its present elementary state, only small scenes with a limited amount of detail can be transmitted, and the frequency of the signal impulses need not rise above the audible frequency limit of about 10,000. For wireless broadcasting purposes the frequency is limited to 4500. Such frequencies can be recorded on a gramophone record without difficulty. If television is to progress and acquire a real entertainment and educational value—as it inevitably will eventually—the technique will have to be very considerably improved to permit of the transmission of much larger scenes, which will have to be capable of reception on a much larger screen than is possible (other than experimentally) at present. When these developments take place the frequency of the transmitted impulses will increase enormously, and it remains to be seen whether the gramophone recording experts will be able to deal with such high frequencies successfully.

Meanwhile, phonovision must remain in its present position, which is that of a scientific curiosity of no commercial value.

Another development with more promising prospects is called "noctovision," which means, as its name implies, seeing in the dark.

During his early experiments with television, using the flood lighting system of transmission, Baird had to use such enormously powerful lights that his sitters experienced very considerable discomfort. Photoelectric cells have been so vastly improved since then that such excessively powerful lights are no longer necessary. Before these improvements were effected, however, it occurred to Baird that he might overcome the difficulty by using rays outside the visible spectrum.

In order fully to understand this development, and the

prospects which have arisen out of it, it might be as well at this point to consider briefly the spectrum.

Readers who are wireless enthusiasts will know that wireless waves are electrical vibrations, the frequency of which depends upon the wavelength. The longer the wavelength the lower the frequency. Wireless waves extend in length from a few centimetres, used only in the laboratory, to nearly 30,000 metres, as used by the great high-power transoceanic wireless telegraph stations. The frequency of vibration of a 1-metre wave is 300,000,000 cycles per second, and that of a 30,000-metre wave is 10,000 per second.

Wireless waves, however, occupy but a small part of the spectrum. Visible light and the various invisible rays are, to-day, accepted as vibrations in the ether, just like wireless waves; only the frequencies at which light and other rays vibrate are very much higher than those of even the shortest wireless waves. A diagrammatic representation of the spectrum, showing all the known frequency bands and the uses to which they are put, is given in Fig. 46.

Of all the component parts, or sections of the spectrum, that section which is most familiar to non-technical members of the public, and at the same time the smallest, is the visible spectrum. This contains the colours, and extends from violet, at the high frequency end of the spectrum, to red at the lower end. As the frequency of the vibrations is reduced, so we pass through the entire range of colours, violet, blue, green, yellow, orange, and red.

Of the entire spectrum, the unaided human senses are only capable of responding to, or detecting, the colours lying between violet and red, i.e. the visible spectrum. To detect the other frequencies, special instruments are necessary, such as, for example, a wireless receiver when it is desired to detect wireless waves, or a fluorescent screen when it is desired to make X-rays or cathode rays visible.

It is interesting to note from the sunlight energy diagram, given in Fig. 46, that the maximum energy reaches us from the sun in the form of infra-red, or heat rays.

In his first attempts to apply invisible rays to his televisor, Baird turned to the upper end of the spectrum and tried ultra-violet rays. He found, however, that although these rays gave some results, they proved unpleasant to sitters. Furthermore, much of the energy contained in ultra-violet rays becomes

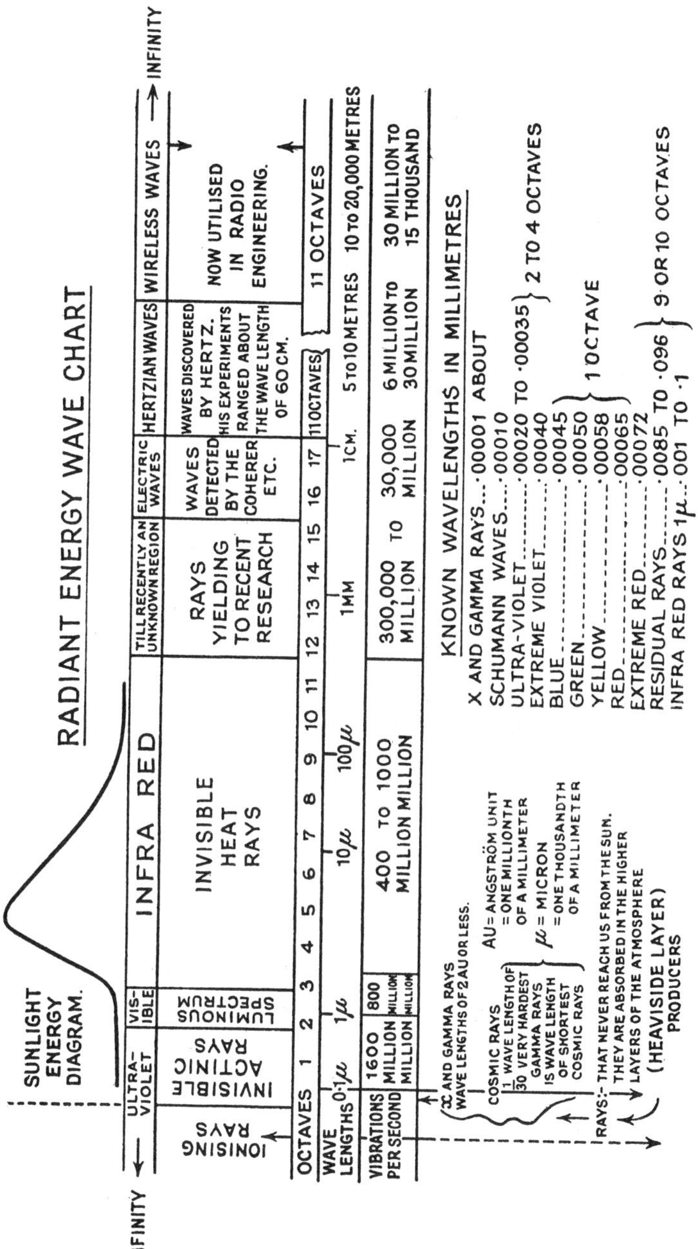

FIG. 46.—The frequency spectrum.

absorbed both by lenses and by the atmosphere. He therefore turned to the other end of the spectrum and used infra-red rays. The potassium photoelectric cells available at that time were chiefly sensitive to violet light, and extremely insensitive to red light and infra-red rays, and this caused difficulties at first. Modern cæsium cells, however, are most responsive to the red end of the spectrum.

Practically all light sources are rich in infra-red rays, and in order to obtain the rays it is only necessary to filter out all visible light with a suitable filter which will allow only infrared rays to pass. Several substances will act as suitable filters, one of the commonest being thin sheet ebonite.

In applying infra-red rays to television in order to produce noctovision, therefore, it is only necessary to cover the light source at the transmitter with ebonite filters. During his early attempts, Baird used the flood-lighting system of transmission; later, he used the spot-light system.

The result produced is extremely uncanny, for the sitter appears to be sitting before the transmitter in complete darkness, and yet he is perfectly visible at the distant receiver.

At first, owing to the imperfections of the photoelectric cells, the received image assumed a perfectly ghastly appearance because, owing to the failure of the cells to respond to red, that colour appeared at the receiver as black. With modern cells, however, there is but little to choose between a television and a noctovision image as regards appearance.

In contrast to ultra-violet rays, which are quickly absorbed by the atmosphere, infra-red rays have extraordinary penetrative powers. Readers will have observed that, in foggy weather, the most intense white lights show dull red through fog. The thicker the fog the duller is the red which shines through. It is for this reason that red neon lamps are used at aerodromes to guide aircraft in foggy weather. The scientific explanation is that the penetrating power of light varies as the fourth power of the wavelength, so that red light penetrates fog some sixteen times better than blue light, and infra-red rays penetrate some sixteen to twenty times better still.

It is evident, therefore, that there is scope for some development of noctovision which will safeguard fogbound ships at sea. With this object in view, Baird experimented with a machine the object of which was to render visible the ordinary navigation lights of a ship, no matter how dense a fog it might

THE BAIRD SYSTEM

be in. This apparatus takes the form shown in the photograph facing page 60, and in the schematic diagram given in Fig. 47.

It consists essentially of a simplified form of combined television transmitter and receiver. The large collecting lens picks up whatever rays are arriving at the apparatus. If the fog is dense enough, distant lights will be invisible to the naked eye and only infra-red rays will reach the lens.

These rays are then focused through the holes in a special scanning disc on to a photoelectric cell specially chosen for its sensitiveness to rays at the red end of the spectrum. After amplification by the amplifier, A, the output impulses from the cell are fed to a second amplifier, B, at the bottom of the instrument.

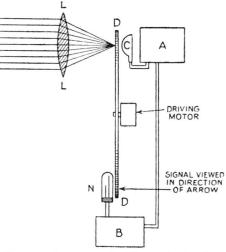

Fig. 47.—Arrangement of the noctovisor for navigational use.

The output of this amplifier is then fed to a neon tube placed behind the scanning disc.

The entire equipment is carefully shielded and housed in a metal box which is mounted on gimbals and a swivel, so that the collecting lens can be swung round in any direction. At the back of the box, near the bottom, is the viewing aperture which the observer watches while searching. Should he pick up a distant light, it will appear somewhere on the screen as a small red spot. By means of two cross wires, this spot can, by swinging the entire instrument, be brought to the exact centre of the screen. This done, the exact bearing of the distant light can then be read off from a graduated scale around which the instrument rotates.

Several types of scanning disc have been experimented with, and one which has proved suitable is shown in Fig. 48. As it is not necessary to produce a finely detailed image of the distant light, but merely to make it visible, holes ⅜ inch in diameter are used. It is desirable to have the holes rather large in order to admit as much of the infra-red ray beam as possible to the photoelectric cell, and thus increase the range of the noctovisor. The diameter of the disc is a matter for some careful consideration, for the neon tube and the photoelectric cell must not be brought into too close proximity or interaction will take place. The neon tube is so connected that it glows only when a distant light has been picked up, so that the light spot on the screen is seen as red on a dark background.

The instrument will operate equally well by day or by night. By day, it is so adjusted that the cell does not respond to an even daylight signal, but only to points of light (or infra-red rays) which have a greater intrinsic value than the steady daylight.

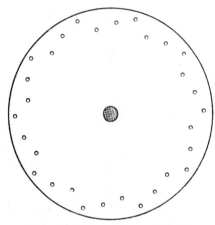

Fig. 48.—Type of disc used in the noctovisor.

It is not difficult to visualise the great value which some development of this idea will be to the captain of a fogbound vessel at sea. By means of it he will be able to pick up otherwise invisible lighthouses, and thus find his way as easily as if he were in clear weather. Similarly, in crowded waters, he could by means of the noctovisor keep a look-out for the navigation lights of other vessels in the vicinity, and thus considerably minimise collision risks.

While television has always, up to the present, been carried out by the use of artificial light to illuminate the sitter at the transmitter, it has been demonstrated by Baird that ordinary daylight can be employed. For the purpose of such demonstrations he uses the flood-lighting type of transmitter, employing a lens disc instead of a Nipkow disc. Sitters placed before such a transmitter in ordinary diffuse daylight (not brilliant sunlight) have been televised with a degree of success equal to that obtained in cases where artificial light is used.

[*Courtesy: Bell Telephone Laboratories.*

Above: The transmitter used by the Bell System in the 1927 demonstration.
Below: Walter S. Gifford, President of the American Telephone and Telegraph Co., using the small screen. Dr. Ives is at right.

THE BAIRD SYSTEM

The possibility of transmitting television in colours intrigued experimenters for a long time before television even in monochrome was possible. Baird has succeeded, however, in demonstrating that colours can be transmitted. The results at the receiver were of the crudest description, as regards recognisable detail in the image, but the colours certainly came through distinctly. The process is a simple three-colour one, in many respects similar to some forms of colour photography and cinematography. It consists in presenting to the eye, in rapid succession, first a blue image, then a red, and then a green image.

These three colours form the well-known primary colours, from combinations of which any other colour or tint may be obtained. For example, purple is a mixture of red and blue ; yellow is a mixture of green and red, and all other colours excepting the three primaries, red, green, and blue, are in a similar way composites of these three colours in varying proportions. When the three colours are combined together they give to the eye an impression of white light.

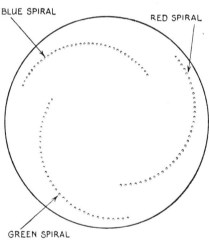

FIG. 49.—The triple spiral disc used by Baird for colour television.

The actual mechanism used in these first colour television experiments consists of a Nipkow disc perforated with three spirals of holes (thirty-six holes in each spiral) arranged consecutively round the disc, as shown in Fig. 49. This disc is used in conjunction with the light spot system of transmission. By using a disc with three sets of perforations it is possible to traverse the subject to be transmitted firstly with a blue light spot, secondly with a red spot, and thirdly with a green spot, the perforations in the disc being covered with blue, red, and green light filters, as shown in Fig. 49.

Due to the absorption of light by the filters, less light is reflected back from the sitter's face to the photoelectric cells than is the case in monochromatic television, so a series of special mirrors is arranged round the sitter's face to reinforce the reflected light. Three banks of photoelectric cells are used, one arranged horizontally on a level with the sitter's forehead, and two vertically, one either side of his face. The cells are carefully chosen so that each bank responds to one of the primary colours. The colour response curves of various photoelectric metals are given in Fig. 50.

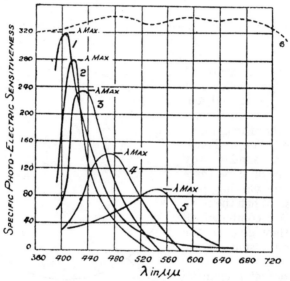

Fig. 50.—Colour sensitiveness of various alkali metals. (1) Lithium; (2) Sodium; (3) Potassium (indigo-blue); (4) Rubidium (green); (5) Cæsium (orange); (6) Colour sensitiveness of a panchromatic photographic plate, which is the ideal to be aimed at in developing a panchromatic photoelectric cell.

The operation of the mechanism, therefore, causes the photoelectric cells to respond firstly to a picture showing only the blue parts of the scene before them; then to a picture showing only the red parts; and lastly to a picture showing only the green parts of the image.

At the receiving end an exactly similar disc is used, with three spirals covered with filters in the same manner as the transmitter disc. The difficulty at the receiving end, however,

THE BAIRD SYSTEM

was to find a light source capable of generating blue, red, and green rays.

The neon lamp which is used for monochromatic reception, while exceedingly rich in red rays, has practically no blue and green components. In order to solve the problem, an attempt was made to construct a special glow-discharge lamp by mixing together neon, helium, and mercury vapour. This experiment proved unsatisfactory, because the blue and red components, and also the green component, varied their proportions, depending upon variable conditions within the lamp.

A solution of the difficulty was ultimately arrived at by making use of two separate lamps, one an ordinary neon lamp, and the other containing helium and mercury vapour, the latter gas providing a very intense blue-green component while the helium provided the blue.

The use of two lamps complicates matters at the receiver, of course, for they must not both be in circuit at once. By means of a commutator arrangement, however, it is arranged that while the red viewing holes of the receiving disc are between the eyes of the observer and the lamps, only the neon lamp is illuminated and controlled by the incoming signal. While the blue and green holes are in the correct viewing position, the commutator disconnects the neon lamp and switches in the mercury and helium lamp. The arrangement of the receiver is shown in Fig. 51.

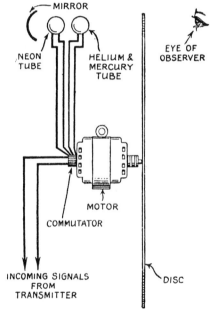

FIG. 51.—Layout of Baird's colour television receiver.

It will be seen that in this arrangement, although there are three separate images, these images are transmitted in succession, and only one channel of communication is necessary for transmission.

On the other hand, the rate of transmission must, theoretically, be increased three times, as there are three times as many images to be transmitted per second. It would appear that unless this is done the three-colour images would not blend properly within the period of persistence of vision. In practice, however, it is found that it is not necessary to increase the speed so much as three times that of monochromatic transmission, for the images have many parts in common, and a much lower speed may be used without introducing a disagreeably noticeable flicker.

By means of this apparatus, Baird succeeded in transmitting images of such brightly coloured objects as flowers, baskets of fruit, gaudily dressed dolls, etc. The three primary colours certainly came through vividly, but the detail left much to the imagination, and intermediate colours could not be distinguished very readily. However, the accomplishment of colour television at all, even in a crude form, is a step in the right direction and a noteworthy achievement.

Still another of Baird's experimental achievements is stereoscopic television. Before proceeding to describe this development, it is, perhaps, desirable to explain briefly a phenomenon of vision which is seldom appreciated, especially by the younger generation who do not remember the stereoscope which was at one time to be found in almost every household as a form of parlour entertainment.

To those of us who still retain the unimpaired use of both eyes, all vision is stereoscopic. That is to say, when we look at any object or scene we gain an impression of stereoscopic relief, depth, or distance, and from such an impression we derive a sense of proportion. A distant house, as seen through the window, may appear no bigger than the flower pot standing on the window table, yet our eyes tell us that the apparent similarity in size is due entirely to the fact that the house is much farther away and is, as an artist would say, foreshortened.

The false impression is, of course, corrected very largely by our experience, which tells us that a house and a flower pot are two totally different things as regards comparative size. But show the same scene to a savage who has never seen either a flower pot or a house, and he will be able to tell you that the house is a much bigger thing than the flower pot.

The savage is able to do this because the use of his two eyes gives him an impression of distance. He can estimate how

THE BAIRD SYSTEM

far away the house is, and by comparison draw the conclusion that the house is a much larger object, even if he has never seen one before. Had he but one eye, however, the picture which he would see would be flat, like a scene on a cinema screen and, having no previous experience to guide him, he would be totally unable to distinguish between the relative sizes of the two objects or form any estimate as to the distance away of the house. Readers should try the experiment of looking at an unfamiliar scene with one eye only.

Normal vision shows us objects in relief, so that we can judge distances and sizes, and tell at once the difference between a photograph of an object and the object itself, in that one is merely a flat picture, whereas the other has depth. This effect of depth is due principally to the fact that our visual impression is a combination of the images seen by our two eyes.

The left eye sees a slightly different view of any object to that seen by the right eye, and the brain combines these two slightly dissimilar views to give us an impression of solidity and depth.

In the once familiar stereoscope, advantage is taken of this principle in order that a photograph may be reproduced in a life-like manner. Two photographs are taken simultaneously by two cameras (combined in one case), the lenses of which are separated by a distance equivalent to the separation of the two eyes, which is about $2\frac{1}{2}$ inches. These two cameras, therefore, produce two photographs of the same object, but from a slightly different point of view.

The finished photographs are then mounted side by side, and viewed through two lenses, mounted about $2\frac{1}{2}$ inches apart, which have the effect of merging the two views into one, giving to the combined picture a natural appearance of depth. By adding colour photography to the process the scenes, viewed in their natural colours, take on a startingly realistic appearance. Perhaps the most valuable application of stereoscopy, however, is in connection with range-finding for artillery or other purposes.

Baird applied exactly the same principles to television, in order to produce a television image in stereoscopic relief at the receiver. Two images are sent alternately, one corresponding to a right-eye view of the scene before the transmitter, and the other corresponding to a left-eye view.

84 FIRST PRINCIPLES OF TELEVISION

At the transmitter a disc containing two separate spirals of holes is used, one spiral being near the periphery, and the other one $2\frac{1}{2}$ inches nearer the centre, so that the two spirals are separated approximately by the distance between two human eyes. The arrangement is shown in Fig. 52, which also illustrates a disc with two sets of triple spirals for obtaining stereoscopic television images in natural colour.

A plan view of the complete transmitter is given in Fig. 53, from which it will be seen that two light beams from two light sources, after passing through the holes of their respective spirals, are focused by lenses L (left-eye view) and R (right-eye view) on to the face of the sitter. The two beams, owing to the arrangement of the spirals, fall on the sitter's face alternately,

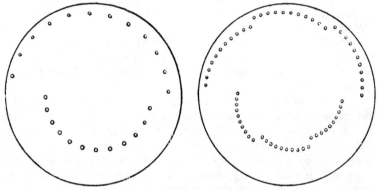

FIG. 52.—Discs used by Baird to achieve television in stereoscopic relief (left) in monochrome (right) in colour.

not simultaneously. Light reflected back from the sitter's face (due to either beam) falls on to the photoelectric cells in the usual manner.

At the receiver a disc geometrically similar to that in use at the transmitter revolves in step with the transmitting disc, so that two images appear side by side, one image corresponding to the object as seen by the left eye, the other corresponding to the object as seen by the right eye.

When viewed through a properly focused stereoscope, these two slightly dissimilar images should combine so that the observer gains the desired impression of stereoscopic relief. A plan view of the receiver is given in Fig. 54, which shows an observer watching the images through a stereoscope.

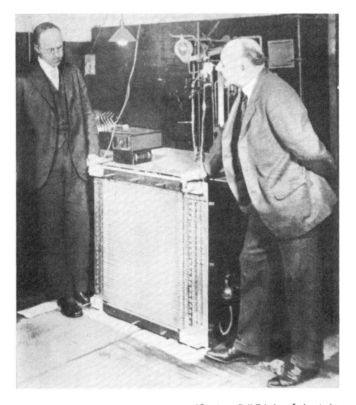

[*Courtesy: Bell Telephone Laboratories.*

Dr. Ives, in charge of the Bell System television work (right), with the large neon tube used in the 1927 large screen demonstrations. The tube is here shown in process of manufacture.

[*To face page* 84.

THE BAIRD SYSTEM

As explained in the last chapter, the size of the area scanned

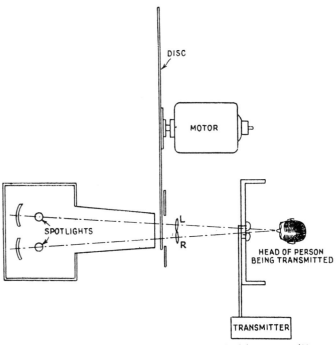

FIG. 53.—Plan view of the stereoscopic television transmitter.

by a Nipkow disc, and therefore the size of the received image, depends upon the radial distance between the outermost and innermost holes of the spiral, and the circumferential distance between successive holes. It follows, therefore, that in order to enlarge the received image it is necessary to enlarge both the diameter of the disc and the size of the neon lamp and its associated amplifier. Most of the equipment described in this chapter gives a received image (unmagnified) measuring about 1 inch wide by about 2 inches high.

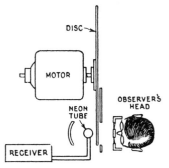

FIG. 54.—Plan view of the stereoscopic television receiver.

Early in 1928, however, by using a lens disc about 8 feet in diameter, Baird produced images at the receiver of a man's head and shoulders, which were life size without magnification. A year later, by using a lens disc and a crater type neon lamp instead of a flat plate type, and projecting beams from the lamp through the lenses of the revolving disc on to a screen, he produced images which measured between 3 and 4 feet across. The illumination of the screen was very poor, however, owing to the low actinic value of the crater neon lamps then available, and the detail also left much to be desired, according to the accounts of those who saw the apparatus in operation. Nevertheless, the experiment was a step in the right direction, for the time must inevitably come when television images will be projected on to full-size cinema screens and be as brilliantly lighted as are our cinema screens to-day. It may be many years before this ideal is reached, or it may be only a matter of months, but at least the need is realised and the first steps have been taken. Particulars of foreign efforts in this direction will be given in later chapters. Details of the commercial Baird " Televisor," which is now on the British market, will also be given in a later chapter.

CHAPTER VIII.

THE BELL SYSTEM.

1927 Monochrome Demonstration. Circuits and Methods Employed. Cells and their Arrangement. Details of Large Screen. Method of Illuminating and Modulating Large Grid Neon. Daylight Demonstration. Direct Scanning. Optical Arrangements. Colour Television. Sodium Cell. Arrangement of Cell Cabinet. Transmitter Amplifier Circuits. Arrangement of Glow Lamps, Filters, and Mirrors at Receiver. Receiver Amplifiers. Adjustment Problems.

FOLLOWING the spectacular successes of Baird in England, one of America's largest industrial concerns and a giant in the world of electrical communications, the American Telephone and Telegraph Co., began to take an interest in television. This company depends for its scientific progress upon vast organised laboratories staffed by some of the cleverest telephone and telegraph engineers in the world.

On April 7th, 1927, a group of engineers attached to these laboratories (the Bell Telephone Laboratories, Inc.) succeeded in giving over a wire circuit between Washington and New York, and over a wireless circuit between Whippany, N.J., and New York, the first demonstration of television ever witnessed in America.

This demonstration was given before a party of guests which included business executives, newspaper editors, engineers, and scientists. The party assembled at the Bell Telephone Laboratories in New York, and were enabled to speak to, and simultaneously see, friends over the ordinary long-distance telephone line to Washington, 200 miles away.

Walter S. Gifford, President of the American Telephone and Telegraph Co., held a conversation with Secretary of Commerce (now President) Hoover, who was in Washington. Mr. Hoover's remarks were made audible to the assembled guests in New York by means of a loud speaker, whilst the image of his face was made visible on two screens, one a small one for individual use, measuring 2 by $2\frac{1}{2}$ inches, and the other measuring 2 feet by 2 feet 6 inches.

88   FIRST PRINCIPLES OF TELEVISION

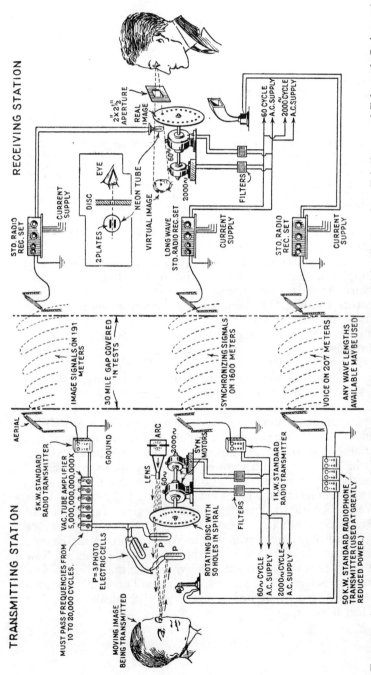

Fig. 55.—A comprehensive diagram of the entire apparatus and circuits used in the American Telephone and Telegraph Co.'s 1927 demonstration of the transmission of simultaneous speech and vision by wireless. Note real and virtual images to left of observer at receiver.

This large screen, which was prepared for the benefit of the visitors, appeared to be somewhat corrugated. This was due to the fact that the squares which made up the picture were arranged in fifty rows, one on top of the other. As the eye became accustomed to looking at the screen in the darkened room the image of the distant speaker's face was recognisable, although the features, which showed up clearly and sharply on the small screen, were considerably blurred by the enlargement and at times disappeared altogether.

It was plain that, enlarged to the size of an ordinary motion picture screen, the detail would have been completely lost. It was stated by those responsible for the demonstration, however, that the invention was far from the picture house stage of perfection; but it is of interest to note that this was the first occasion on which a large screen was ever demonstrated.

The images reproduced on the small screen were fairly clear, however, and were described by witnesses as being comparable to faded daguerreotypes held in a shaky hand. In these small pictures the detail of the face appeared in black lines against a shining gold background, which latter was due to the orange light from the neon tube which was used as the light source.

Following the wire demonstration between New York and Washington, the visitors were entertained to a repetition of the experiments, the transmission on this occasion being by wireless from the company's experimental station, 3XN, at Whippany, N.J., about thirty miles from New York. An important feature of the demonstrations was that there was no difference in the quality of the reproduced image, whether it was transmitted from one end of the laboratory to the other, by wire from Washington, or by wireless from Whippany.

The engineers responsible for these demonstrations made it quite clear that their apparatus and methods could not yet be brought into regular commercial use. The necessary equipment was extremely elaborate and costly, and for the purpose of the public demonstration the services of a total of nearly 1000 men were required! However, it is characteristic of our friends across the Atlantic that they frequently have to resort to exaggeration in order to impress their fellow-countrymen.

Another feature of the experiments was that a wide frequency band was required for the transmissions, which necessitated, in the case of the wire transmission, the use of several telephone circuits. In the case of the wireless transmissions, one wavelength had to be used for sending the picture impulses, one for the synchronising impulses, and a third for speech, as shown in Fig. 55. In view of the already seriously congested state of the ether, this presents a formidable difficulty to commercialisation. This question of frequency bands will be discussed more fully in a later chapter.

A comprehensive chart of the wire circuits used between Washington and New York is given in Fig. 56. Underground cables were used for the programme (speech), synchronising, and order circuits. For the television signals open wire lines were used, except through cities and under rivers, where underground cables were employed, as indicated by the thick shaded sections in Fig. 56.

Much of the knowledge used by the engineers responsible for the development of the system of television under review was gained by them during the course of years of development and research in the field of phototelegraphy as a result of which the A.T. & T. now operates on a commercial basis a highly efficient system of still picture transmission by wire which covers the entire United States. Television and phototelegraphy, although, as we have already seen, two totally different things, have several points in common, or at least closely associated, so that the engineers who developed the phototelegraphy system were able to make use of much of the information they had gained.

The system of transmission employed by the Bell Telephone Laboratories is identical in principle with that already described in the chapters dealing with the Baird system, viz. the Nipkow disc or light-spot system, but usually referred to by Bell engineers as the beam scanning method. The main point of difference between the Baird and the Bell methods lies in the fact that in the latter case the scanning beam is directed through the top of the disc instead of through the side, as is done by Baird, thus achieving horizontal instead of vertical scanning. The respective merits of these two methods will be discussed in a later chapter. Inasmuch as the resources of the Bell Laboratories, both in money and in scientifically trained men, far exceed those of Baird, their apparatus is better in design and finish, is far more costly and much more complicated.

A steel scanning disc drilled with fifty holes was employed, and the light source at the transmitter was a Sperry arc lamp

THE BELL SYSTEM

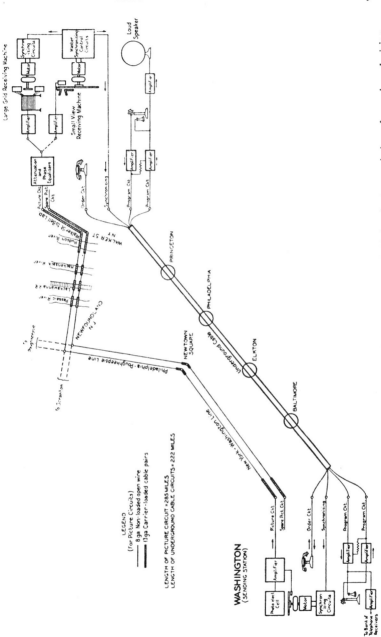

FIG. 56.—Comprehensive chart of the wire circuits used by the A. T. & T. for the transmission of speech and vision from Washington to New York.

consuming 40 amperes. Such a powerful light source naturally provided an intense illumination, as was necessary in view of the insensitiveness of the photoelectric cells then available. However, due to the extremely transitory nature of the illumination due to the travelling light-spot, the light was not altogether unbearable, although it was stated that sitters began to feel a sensation of discomfort after a few minutes' exposure to it.

As we have already seen, light from the sitter's face is diffusely reflected in all directions, and in the case of the Bell system some of the light which is reflected forward falls on three large photoelectric cells which are connected in parallel and operate through a common amplifier system.

In contrast to the flood-light system of transmission, the light-spot method permits two very large gains in the amount of light which can be collected and transformed into photoelectric currents. In the first place the transient nature of the illumination permits a very intense illumination to be used without inconvenience to the sitter. In the second place the optical efficiency of the system is not limited to the aperture of a lens used in conjunction with a single small photoelectric cell, but can be increased by using large cells and more than one cell connected in parallel.

The cells used in the present case are probably the largest ever made. They measured about 14 inches long, and each cell represented an area of 40 square inches of photoelectric surface. One cell was arranged horizontally and slightly above the sitter's face whilst two others were arranged vertically, one on each side of the sitter.

The lights and shadows of a television image are the same as if the light came out of the three large photoelectric cells to illuminate the sitter. As a consequence the problem of lighting the sitter becomes a problem not of manipulating light sources but of manipulating the photoelectric cells, and the considerations governing their disposition are the same as those which govern the disposition of lights in a photographer's studio. If the effect of lighting from one side is desired, there should be more cells on that side. If the effect of diffused lighting is desired, a large number of cells, or cells of large area, should be disposed all around the subject. Shadows are not produced where no light falls on the sitter, but at those points from which no light can reach the cells.

[*Courtesy*: *G.E.C.* (*U.S.A.*).

Dr. E. F. W. Alexanderson, of the General Electric Company (U.S.A), and his mirror-drum seven-light-spot projector with which he experimented in 1927. By using seven spots and seven communication channels he hoped (and still hopes) to obtain greater detail and illumination.

[*To face page* 92.

THE BELL SYSTEM 93

In contrast to this American statement of the case, it is interesting to note that Baird uses four small cells grouped as closely together as possible and placed in front, and slightly above the face of the sitter. He places no cells either at the sides or below the sitter's face.

The scanning disc used in the Bell system revolved at 1080 R.P.M., which means that the scene to be transmitted was scanned eighteen times per second. Operating at this speed it was stated that the television signals reached a frequency as high as 20,000 per second. From the data given, this figure would appear to be correct by calculation, but, as will be explained in greater detail in a later chapter, was probably not reached in practice.

As already mentioned, two forms of receiving apparatus were used at the demonstration. In each form a neon tube was employed. In the small receiver a neon lamp of the flat plate type was used, but for the presentation of the image to a large audience a different method was tried experimentally. Instead of using a single relatively small neon tube, successive portions of which are viewed at successive intervals, a very long tube was used, which was bent back and forth upon itself to form a grid, as shown in Fig. 57, which illustrates the arrangement diagrammatically.

The tube is bent into fifty loops, corresponding to the number of holes in the scanning disc at the transmitter. Instead of fitting this long tube with a single pair of electrodes, as is done in the case of the smaller tubes, the grid is equipped with 2500 electrodes, fifty per turn of the loop. These electrodes consist of pieces of metal foil cemented on to the outside of the tube. The interior electrode (anode) is a long spiral of wire. Each of the exterior electrodes corresponds to a single elemental area of the picture plane which is scanned by the photoelectric cells at the transmitting end.

These electrodes are connected by individual wires to a distributor, or commutator, which in turn is connected to the amplifier which handles the incoming television impulses. The distributor revolves in exact synchronism with the scanning disc at the transmitting end.

When a particular spot on the object being transmitted is illuminated, its position and light intensity are transmitted in the form of an electrical impulse to the receiving station, as has already been described. In this case, however, the distributor

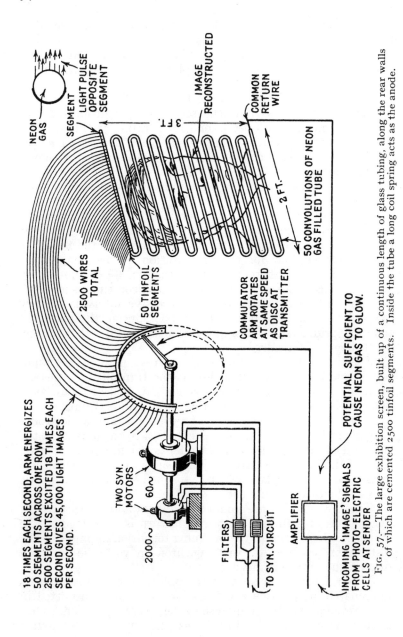

Fig. 57.—The large exhibition screen, built up of a continuous length of glass tubing, along the rear walls of which are cemented 2500 tinfoil segments. Inside the tube a long coil spring acts as the anode.

selects the proper connection (cathode) of the neon tube grid and lights a spot on it which corresponds with the spot on the original scene, the illumination of which set up the impulse in the first place. Like the neon tube the distributor has 2500 contacts, or commutator segments, and 2500 wires are thus necessary to connect the segments to the neon cathodes.

The method of illuminating the large grid neon is very different from that adopted in the case of the smaller flat plate type. This is necessary because the impedance to which energy must be supplied differs materially, due mainly to capacity effects.[1] For low voltages the impedance between any electrode (cathode) and the central helix (anode) is effectively a capacity of the order of 6 mmfd. When, however, the voltage gradient inside the tube rises sufficiently to cause the gas to break down and start a glow discharge, the capacity is increased to about 15 mmfd. In fact, the tube may be looked upon as consisting of two capacities connected in series. When the applied potential is sufficient to cause a discharge, that capacity corresponding to the portion of the path inside the tube is effectively shunted by an ohmic resistance consisting of the now glowing gas.

The minimum discharge potential has been found to be independent of frequency over a wide range, but the current between electrodes is inversely proportional to the frequency because of the presence of the capacity between the electrode (cathode) and the glowing gas. Now, the brightness of the discharge is a function of the current sustaining it, so that it becomes desirable to use high frequencies in order to get sufficient light without going to prohibitively high voltages. It is also desirable to operate at such a portion of the frequency scale that the percentage difference between the limits of the range shall be small, thus avoiding signal distortion due to the effect referred to above. There is, however, a definite upper limit to the frequency beyond which it would be impossible to operate because of the stray capacities in the cable connecting the grid to the distributor. It has been found feasible to operate at a frequency of 500,000 cycles (600 metres).

The solution to the problem was conveniently found by making use of a radio broadcast transmitter arranged to operate at that frequency. The voice frequency circuits of the trans-

[1] The following description of grid neon operation is taken from the *Bell System Tech. Journal*, Vol. VI., Oct. 1927.

mitter were suitably modified so that the extended range of frequencies required could be handled with minimum distortion.

The envelope of the 500 Kc. wave modulated by the picture signal, as shown in Fig. 58, is proportional to the signal amplitude plus a direct current biasing component of such magnitude that when the envelope drops to 160 volts the neon tube fails to glow. This corresponds to a black area in the picture.

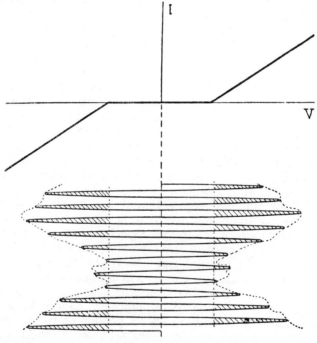

FIG. 58.—Showing the relationship between the modulated high-frequency wave impressed on the grid type neon tube and the neon tube characteristics. The intensity of the glow is proportional to the shaded area.

When no picture signal is being received, the amplitude of the unmodulated carrier wave causes the tube to light at average brightness, corresponding to the locally introduced D.C. component of the signal. It follows that the amplitude of the unmodulated carrier is fixed by the joint requirements of two biases, that of the lamp (to cause the initial glow) and that of the signal bias.

It should be explained here that only the alternating current

Above: Spotlight transmitter experimented with by the G.E.C. (U.S.A.) in 1928. The bulbous projections in front of the subject are the photoelectric cells. The object resembling a miner's lamp is the microphone. *Below:* An American experimental television receiver, using horizontal scanning

[*To face page* 96.

variations produced by light changes are amplified and transmitted from transmitter to receiver. Any steady current component output of the photoelectric cells, due perhaps to the steady illumination of the room in which the sitter and the cells are situated, is not sent, mainly because of the difficulty of amplifying direct currents. At the receiver this deficiency is made up by introducing a D.C. bias upon which is superimposed the incoming A.C. signal currents. This bias is arbitrary and can be varied to suit circumstances or taste. For instance, by reducing the signal bias the received image can be made to appear dim and soft, lacking hard contrast. By increasing the bias the image can be made very bright, with extremely harsh contrasts.

There is a slight distortion inherent in this high-frequency method of supplying power to the grid neon, due to the fact that the light (which is proportional to the shaded area of the curve in Fig. 58) is not strictly proportional to the amplitude of the envelope with respect to the 160-volt limit. This is, of course, because these peaks are portions of a sine wave and hence the time variation of the glow resulting from any given carrier cycle is a function of its amplitude. The effect is small, however, being most noticeable at low values of illumination.

In the case of the grid neon receiver the signal amplitude is adjusted, as in the case of a flat plate neon and disc receiver, by a potentiometer in the low frequency part of the circuit, i.e. in the modulation circuit of the adapted radio transmitter. The carrier amplitude, however, is adjusted by varying the potential applied to the anode of the oscillating valve. The coupling to the grid neon is made by connecting the central helix (anode) and the revolving brush of the distributor across a portion of the condenser of the oscillating circuit.

The frequency-amplitude relation of the envelope has been made practically constant by employing resistance-capacity coupling in the signal input amplifiers, by providing extremely high inductance retard coils for the modulator—which is of the Heising type—and by inserting resistance in the oscillating circuit to provide sufficient damping. The relations between the original picture signal and the envelope of the high-frequency wave, with respect to both amplitude and phase shift, were observed over the signal frequency range by means of a Braun (cathode ray) tube and found to be satisfactory. The impedance of the connecting leads to the commutator was also

98   FIRST PRINCIPLES OF TELEVISION

measured and found to have a negligible effect on the frequency and damping of the oscillating circuit.

It has been found that there may be a lag between the time when the potential is applied to an electrode and the time when the gas breaks down. This is especially true, following an interval during which there has been no discharge within the tube. Because of this those electrodes which are the first to be connected in any one of the parallel portions of the tube may fail to light. To overcome this effect a small pilot electrode is kept glowing at the left-hand end of each tube, thus irradiating the branch in such a way that the illumination of all electrodes

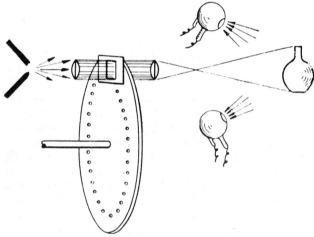

FIG. 59.—The arrangement of apparatus for beam scanning. Light from a single source is projected as a small moving spot on the subject. The reflected light is received by several photoelectric cells.

follows immediately upon the application of potential. These pilot electrodes, which are supplied by means of an auxiliary connection to the oscillator with a potential somewhat lower than that ordinarily impressed upon the picture segments.

By the middle of 1928 the researches of the Bell Telephone Laboratories had progressed sufficiently to enable them to demonstrate television, using ordinary daylight instead of artificial light to illuminate the subject before the transmitter. Baird had already accomplished this feat but, as we have already seen, by means of a lens disc instead of a Nipkow disc. The Bell engineers accomplished the same end by the use of the same Nipkow disc which they used in their 1927 demonstrations, but instead of using beam scanning they employed what they term direct scanning.

The difference between the spot-light and flood-lighting systems of transmission has already been explained in previous chapters. It has also been written that all optical systems are reversible, and this is very clearly demonstrated in Figs. 59 and 60, the first of which illustrates spot-light (beam scanning) transmission, and the second flood-lighting (direct scanning) transmission, using a Nipkow disc instead of a lens disc.

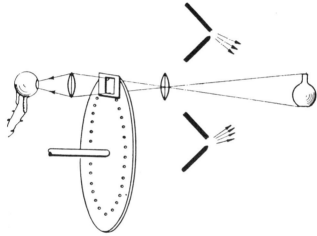

FIG. 60.—The reverse method—direct scanning. Several light sources illuminate the subject : a lens forms an image of the subject on the disc, and the image is scanned by a spiral of apertures through which light falls on a single photoelectric cell.

One of the conditions which made it possible to use ordinary daylight in conjunction with the system of scanning, shown in Fig. 60, was a considerable improvement in the sensitiveness of photoelectric cells. By increasing both the diameter of the disc and the diameter of the holes (while keeping the number of holes the same), more light was admitted to the cell through the disc apertures.

The actual apparatus employed consisted of a 50-hole disc 3 feet in diameter, upon the surface of which an image of the scene was formed by an $f/2 \cdot 5$ lens. By means of this arrangement the image of a full-length human figure standing

30 feet from the transmitter could be televised. The entire transmitter was set up on the roof of the laboratories, ordinary daylight being used to illuminate the subject instead of the arc lamps shown in Fig. 60.

As in photography, the distance of the subject from the transmitter, and how much of the subject is seen, depends upon the kind of lens employed. In the apparatus under discussion it was not possible to scan a full-length human figure at a distance of less than 30 feet because the type of lens employed, if it had been of shorter focal length, would not have covered the whole scanning field on the disc. On the other hand, if the figure to be scanned is farther away it is possible, just as in photography, either to move closer to the subject or use a longer focus lens of the same ratio of diameter to focal length.

It will be by means of the latter expedient that it will one day be possible, with several different lenses, to televise both close-ups and distant views of outdoor events. It could be done to-day, after a fashion, but with present-day apparatus the detail of distant views would be so poor that scarcely anything would be recognisable.

In July, 1929, the Bell Telephone Laboratories successfully demonstrated television in colours. Like Baird, they utilised the principles employed in three-colour photography, but their method of applying these principles was different.

Instead of using a triple spiral scanning disc with each spiral of holes covered with appropriate filters, and one communication channel, as Baird did, the Bell engineers used their usual 50-hole scanning disc, beam scanning, three sets of photoelectric cells and filters, and three communication channels, one for each colour.

Before television in colours could be achieved by this method the first essential was a photoelectric cell which would be as nearly as possible equally sensitive throughout the entire visible spectrum. A cell conforming to these requirements was developed, having as its sensitive surface the alkali metal sodium, which was given a special treatment of sulphur vapour and oxygen. The wavelength response curve of one of these cells is given in Fig. 61, on which is also plotted, for comparison purposes, the characteristic curve of an ordinary potassium hydride cell. The extension of sensitiveness to the red end of the spectrum in the case of the special sodium cell is noteworthy.

As in 3-colour photography it is necessary, when using these cells, to employ suitable colour filters which are placed immediately in front of the cells. In all, twenty-four cells were used, arranged on either side of and slightly above the face of the subject, as illustrated in Fig. 62. The arrangement of the filters, and the number of them for each primary colour, was adjusted so as to give approximately equal response to all three colours. For the blue part of the spectrum two potassium cells having a characteristic similar to the potassium curve in Fig. 61 were used. Eight sodium cells covered by green filters dealt with

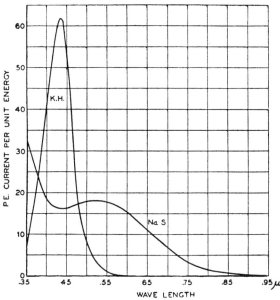

Fig. 61.—Characteristic curve of potassium hydride, and of special colour sensitive sodium (NaS) cells used by the Bell Telephone Laboratories for television in colour.

the green part of the spectrum, while fourteen sodium cells covered by red filters handled the red portions of the image. The total light-gathering power of the cells was augmented by five glass mirrors arranged as shown in Fig. 62.

All the cells were mounted in metal cases lined with sponge rubber, and the filters were carried in light wooden frames which slid into grooves in front of the cells. In order to improve the mixing of the three colours diffusing (frosted) glasses were slid into slots placed some distance from the cells, as indicated in Fig. 62. With the exception of this arrangement of cells the transmitting apparatus was identical with that already described in the early part of this chapter.

The output currents from the cells had to be amplified approximately 10^{13} times before they could be fed to the transmission channels. To avoid interference not only from outside disturbances, both acoustic and electric, and from inter-action between one colour channel and another, a novel arrangement of the amplifying system was resorted to.

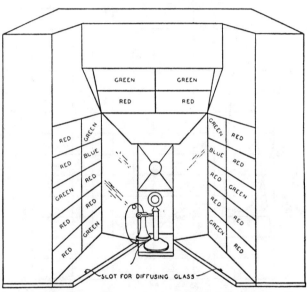

FIG. 62.—Showing the arrangement of photoelectric cells for three-colour television, as seen by the subject sitting before the scanning beam which issues from the circular opening in the centre.

The signals from the cells were first amplified in close proximity to the cells by a two-stage amplifier in each colour channel, these amplifiers being placed within the photoelectric cell cabinet. From these amplifiers three shielded cables carried the signals to a triple 7-stage amplifier. These three amplifiers were arranged side by side on a vertical panel in such a way that the valves were mounted in zig-zag positions in each channel to make the grid leads extremely short, and adjacent stages in all three channels carried approximately equal signal strength to avoid both mutual and stage-to-stage interference.

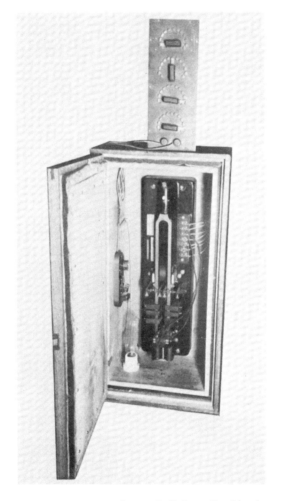

[*Courtesy: Radio Corporation of America.*

Electrically driven tuning-fork, in constant temperature box, used by the Radio Corporation of America for synchronising transatlantic phototelegraphy.

THE BELL SYSTEM

After having been amplified to a convenient level the signals in each channel were fed through permalloy-cored transformers into the three transmission lines. As in monochrome transmission, it was necessary for each of these channels to be capable of transmitting a range of frequencies varying from 18 to 20,000 cycles per second.

At the receiving end the same 50-hole disc was used as for monochrome transmission. Behind the disc were placed three special glow lamps, a neon lamp for the red image and two argon lamps for the green and blue images. All these lamps were of the flat plate type previously described, the only difference being that the plates were made long and narrow and the lamps were so placed as to be viewed greatly foreshortened so as to increase the apparent brightness. This arrangement was made

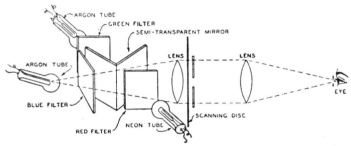

FIG. 63.—The arrangement of lamps, colour filters and mirrors at the receiving end.

necessary because the efficiency (i.e. ratio of light intensity to input current) of argon is low by comparison with neon. In addition, the cathodes were given a hollow tubular form so that a current of water could be sent through for cooling purposes, thus permitting larger input currents to be used to further increase the brilliancy.

The image of the cathode was focused upon the pupil of the eye of the observer by means of a pair of convex lenses, as shown in Fig. 63. The lens nearer the eye was mounted a considerable distance away from the disc for the purpose of magnifying the image. In order to make it possible to see all three lamps simultaneously a pair of semi-transparent mirrors inclined at an angle of 45° were employed. The mirror for reflecting the neon lamp was of clear glass, while that for reflecting one of the argon lamps was thinly silvered to about 50 per cent. reflecting power. In front of the neon lamp a deep red filter (Wratten, No. 29) was used, providing monochromatic red light of equivalent wavelength of approximately ·64μ. In front of the two argon lamps green and blue filters were used (Wratten, No. 7 + No. 44 and No. 47 respectively), the lines transmitted giving light of equivalent wavelength ·53μ and ·46μ, these being substantially the primaries for which the sending end photoelectric cells and filters were adapted.

The amplifier system for controlling the receiving lamps was designed to supply the relatively large currents made necessary by the low efficiency of the argon lamps and the sacrifices of light in the optical system shown in Fig. 63. The picture signals coming in at the receiving end were amplified for each colour by means of a 3-stage amplifier, the last stage employing a 5-watt valve. The output of these amplifiers was passed through special low resistance 250-watt valves in the plate circuits of which the glow lamps were connected. For the green signal two 250-watt valves were used in parallel because of the relative lack of green radiation from the argon lamp. Six potentiometers were provided by means of which the direct current (bias) through each glow lamp and the signal current coming from the transmission line could be separately controlled.

The method of adjustment of this three-colour receiver is interesting.

The first adjustment consists in regulating the three lamps so that their superimposed light appears to be white in colour. This is done partly by varying the amount of the direct current upon which the alternating signal current from the transmitter is to be superimposed, and partly by selecting mirrors of different reflecting powers, or filters of different density. A good white light should be obtained with the lamp currents at a convenient value with respect to the range of adjustment. The desired condition having been attained, the three-colour signals are then received from the transmitter, using a black and white test object, and the three signal currents are adjusted until the image appears black and white. When this adjustment is made a coloured object may be put in position before the transmitter and will appear approximately correct in colour at the receiver.

In all systems of television transmission at present in use, the direct current component of the signal, on which the general tone of the picture depends, is not transmitted, but is applied

THE BELL SYSTEM

at the receiving end in the form of a direct current bias to the receiving lamp. This introduces an interesting complication in the case of colour television which is illustrated by imagining that the black and white test object is replaced by a uniform coloured field. Since there is no alternating current signal now being transmitted, the receiving end field will still appear white.

Similarly if, when coloured objects are to be transmitted, the average colour of the whole field is altered, an adjustment must be made in the direct current components. An extreme case is presented by an object which is pure red, green, or blue, so that only one set of cells sends signals. A green object on a black background will be rendered as an unsaturated green on a purple background, until the red and blue direct currents are switched off. Consequently, the only adjustment which need be made as different coloured objects are introduced, once the black and white object has been properly adjusted for, is in the direct current controls at the receiving end, and these, for the reason indicated, are much more important than in monochrome television, where the general tone may be either light grey or harsh black and white, according to the taste of the observer at the receiving end.

Certain characteristic sources of unfaithful colour rendering deserve special mention. The Bell engineers found that in general the rendering of tone values and of image details must be far more accurate for colour television than for monochrome work. If, for instance, in any one colour the image is being delivered from a curved portion of an amplifier valve characteristic, but not in the other two, spurious colours will be introduced in highlights or shadows. Similarly, if one image contains more marked transients or degradation of edge sharpness than another (which may be caused either by small differences in the electrical circuits, or by different characteristics of the glow lamps), colour fringes result. Another source of error occurs with objects which are scanned in the round, such as the human face, if the diffusion of light to the photoelectric cells is insufficient. In this case the image shows coloured shadows and highlights exactly as if the object were illuminated by several differently coloured lamps whose light is not adequately mixed before falling on the subject.

According to a Bell Telephone Laboratories publication, upon which the above description of colour television is based, the coloured images produced by this method of three-colour television are " quite striking in appearance," in spite of the rather low brightness and small size which are characteristic of the present stage of development. It is stated that the addition of colour contributes notably to the naturalness of the images, and constitutes, from a technical point of view, a considerable advance. " It is obvious, however," continues the Bell statement, " from the discussion which has been given of the problems encountered, that television in colour is intrinsically far more complicated and costly than television in monochrome, and hence is likely to wait much longer for practical utilisation."

107

CHAPTER IX.

METHODS OF SYNCHRONISM.

Independent Control. Synchronising in Phototelegraphy. Electrically-Driven Tuning-Forks. Crystal-Controlled Valve Oscillators. Direct Control. Synchronous Motors and Phonic Wheels. Defects Caused by Lack of Synchronism. Isochronism and Synchronism Defined. Hunting Propensities of Motors. Applying Synchronous Motors to Check Hunting. Method of Phasing the Image. Transmission, Amplification, and Application of Isochronising Impulses. Using A.C. Mains.

THUS far we have neglected entirely one of the most vital problems of television, that of securing and maintaining synchronism between the transmitting and receiving mechanisms. The time has now come when we must examine this phase of our subject in detail.

In phototelegraphy, as distinct from television, the problem of synchronism is relatively simple of solution, for the speed of transmission is so very much slower. Also, as will be made clear later, synchronism in phototelegraphy need only be carried out in what might be described as a single dimension, as far as the automatic mechanism is concerned; whereas, in television, it must be carried out, in a sense, to two dimensions.

In most systems of phototelegraphy the problem resolves itself into the rotation at precisely equal speeds of two cylinders, one at the transmitter and one at the receiver. The recording device at the receiver can be set to the correct starting-point by hand, and simultaneous starting of the two cylinders can be effected by some form of prearranged signal.

The maintenance of equal speeds of operation of different mechanisms can be accomplished in several ways which have been devised from time to time for different purposes. The methods employed may be roughly divided into two classes, in one of which independent generators of a controlling current of constant frequency are used at both the transmitting and receiving stations. In the other class a constant frequency current, generated at the transmitter, is sent to the receiver and applied there to control the speed of the mechanism.

Devices of the former class generally make use of electrically driven tuning-forks or clockwork-operated pendulums as the means of generating accurately timed impulses. For slow-speed work it is sufficient to arrange for a swinging pendulum to close a contact at the end of its swing, but for higher speed work, or for greater accuracy, it is better to make use of the more rapid impulses obtained from a vibrating fork.

The general arrangement of an electrically operated tuning-fork is shown in Fig. 64. Current flowing through the windings of the electromagnet attracts the fork, pulling it over so that the interrupter contact is opened, thus switching off the current and permitting the fork to return to its normal position, when

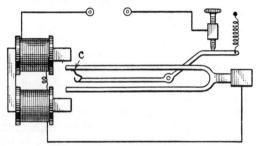

FIG. 64.—A form of tuning-fork interrupter which is kept in vibration by means of an electromagnet.

the contact once more closes and the fork is again attracted. The action is similar to that of a buzzer, only, in this case, the perfectly tuned fork (which acts as an armature) sets up, in the battery circuit, an intermittent current of regular periodicity. The frequency of the current corresponds exactly with the natural period, or frequency of vibration of the fork.

Representative of the manner in which a fork can be applied to synchronise two remotely separated mechanisms is the system of synchronism developed by Captain R. H. Ranger, of the Radio Corporation of America, and used in the transmission of photographs, facsimiles, etc., between London and New York, and between New York and Honolulu, by combined wire and wireless transmission. These transmissions, it should be clearly understood, constitute phototelegraphy, not television.

The circuit arrangements of the Ranger synchronising

[*Courtesy: G.E.C. (U.S.A.).*

A simple form of synchronous motor.

[*To face page* 108.

METHODS OF SYNCHRONISM

method are given in Fig. 65. The heart of the device is the electrically operated tuning-fork, which is mounted in an insulated case. This case is kept heated to a constant temperature by means of a lamp. By keeping the temperature constant, variations in the frequency of vibration of the fork, due to expansion and contraction at various temperatures, is avoided.

The tuning-fork is caused to vibrate at a frequency of 70 per second by means of the electromagnet and interrupter contacts shown, and a chronometer, operating through a relay and the correcting magnet, keeps the frequency of vibration constant.

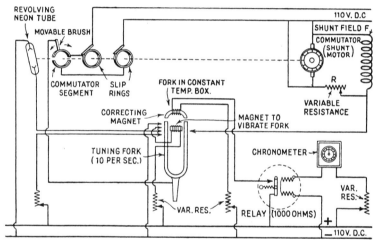

Fig. 65.—Diagrammatic sketch of the tuning-fork controlled synchronising mechanism developed by the Radio Corporation of America.

In the upper right-hand corner of Fig. 65 is the D.C. shunt-wound motor which drives the photo-transmission mechanism, and the vibrating fork keeps its speed of rotation constant in the following manner.

If, at a given moment, all the tuning-fork and commutator contacts are closed, the variable resistance in series with the field winding of the motor is short-circuited and the speed of the motor is reduced, due to the sudden rush of current through the field winding.

At another moment it may be that the contacts are closed through the tuning-fork and auxiliary commutator segments and slip rings at the left-hand end of the motor shaft. In such a case the field winding of the motor will be short-circuited, and due to the greatly weakened field the speed of the motor will increase. This action goes on repeatedly, so that the *average* speed of the motor remains constant.

The motor is designed to run at a speed of 2100 revolutions per minute. In order to check the synchronism a neon tube is fitted on to the end of the motor shaft, and revolves with it. This tube receives an impulse for each vibration of the fork, i.e. 70 per second, or 4200 per minute. The tube therefore receives two impulses for each revolution of the motor shaft. The result, to the observer, is a stroboscopic effect. That is to say, if the motor is revolving in perfect synchronism with the tuning-fork, the neon tube appears to be stationary. If the speed is below synchronism, the neon tube will appear to be turning slowly backwards, like the wheels of a motor car often appear to be on the cinematograph screen. If the motor speed is above synchronism, then the neon tube will appear to turn slowly forward.

Exactly similar driving motors and tuning-forks are supplied to the transmitting and receiving stations, particular care being taken over the tuning of the fork, so that the transmitting and receiving mechanisms are certain to revolve at the same speed, i.e. in synchronism.

This method, although eminently satisfactory for photo-telegraphy, is not suitable for television, for it is rather complicated, and with the greatly increased speed of transmission involved in television the degree of accuracy obtainable is not sufficiently great. There is also the difficulty that the action of such mechanisms is similar to that of the governor of a steam engine. That is to say, until a variation of speed *has actually taken place* the corrective action does not take place. Thus, although the *average* speed is constant, the speed from moment to moment may vary by quite a fair amount. For television purposes the ideal synchronising mechanism must keep the transmitting and receiving mechanisms at an absolutely dead constant speed at all times.

In another system for securing independent control a valve oscillator, controlled by a piezo-electric crystal, is used in conjunction with a synchronous A.C. motor which is coupled to and controls the main driving motor. This is more accurate and constant in operation than the tuning-fork method, and

METHODS OF SYNCHRONISM

good results have been obtained with it in television work. As may well be imagined, however, such an arrangement is both expensive and complicated, and great care is necessary to preserve accuracy of operation.

Some idea of the accuracy required for television purposes may be gained from the statement made by three American workers, Messrs. Weinberger, Rodwin, and Smith,[1] to the effect that for successful television an accuracy in synchronising frequency of *one part in seven millions* is necessary. An example of such a degree of accuracy would be a clock which, in three months, gained or lost no more than one second. The greatest accuracy which has so far been secured from piezo-electric crystals is only about one part in five millions.

The other class of synchronising method, in which the synchronism of the receiver mechanism is under the direct control of a constant frequency generator at the transmitter, involves the transmission to the receiver of the constant frequency current, and the use of a control mechanism at the receiver which is capable of being operated at a speed which is dependent solely and absolutely on the frequency of the incoming impulses.

It matters little what form of constant frequency generator is employed at the transmitter, so long as the generated frequency is sufficiently high. It may be an electrically operated tuning-fork, or a valve oscillator controlled by a tuning-fork or a quartz crystal; or it may be a small alternating current dynamo coupled to the shaft of the main driving motor of the transmitter.

At the receiver it is usual to cause the incoming constant frequency current to drive a small synchronous motor which is coupled to the shaft of the main receiver driving motor. Synchronous motors consist essentially of an armature, or rotor, supplied with an alternating current; and a stator supplied with direct current. Or in an even simpler form the rotor may be a toothed iron wheel or a wooden drum with longitudinal iron strips which is acted upon, or driven round by a series of electromagnets (stator) supplied with A.C. Such a simple form of synchronous motor is frequently referred to as a phonic drum, or wheel, and is illustrated in Fig. 66. Each A.C. impulse energises the stator magnets and pulls the rotor round, due to the attraction between the magnets and the iron teeth or strips of the rotor.

[1] *Proc. I.R.E.*, Vol. 17, No. 9, p. 1590.

The speed at which such motors run is entirely dependent upon the periodicity, or frequency, of the alternating current supply, and upon the number of poles present in the rotor or stator, whichever is receiving the A.C. supply. It follows, therefore, that if the periodicity of the A.C. supply is absolutely constant, the speed of rotation of the synchronous motor is also constant. Furthermore, if such a motor is coupled to another motor which is inclined (if left to itself) to vary in speed slightly, the synchronous motor will exert sufficient force to keep the second motor running at a constant speed also.

Thus, by generating a constant frequency current at the transmitter, causing part of it to control the speed of rotation of the transmitter driving motor, and sending part of it over a

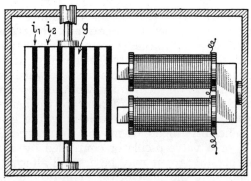

FIG. 66.—The simplest form of synchronous motor, known as a phonic drum or wheel.

circuit to the receiver, there to drive a synchronous motor, the receiving mechanism can be kept running at exactly the same speed as that at the transmitter.

The above method gives more accurate control over the two mechanisms than does the previous method of independent control, but it has the disadvantage that a separate channel of communication is necessary over which to send the synchronising impulses. With independent control absolute constancy of the frequency of the controlling current is a *sine qua non*, but when the direct control method is used constancy is not a first requisite provided that the method of applying the control is sufficiently positive, for whatever variations affect the transmitter, they affect the receiver in like degree and absolutely in unison.

METHODS OF SYNCHRONISM

The system can thus be simplified by dispensing with an elaborate constant frequency generator. In place of it, a small A.C. dynamo can be mechanically coupled to the shaft of the motor which drives the transmitter mechanism, the generated current being sent to the receiver, there to drive the synchronous motor. It then follows that if the transmitter motor varies slightly in speed, the frequency of the A.C. generator will vary, which in turn will cause the speed of the synchronous motor and the receiver driving motor coupled to it to vary. The fact that they do vary slightly does not very much matter, for *they vary in exact unison*, and synchronism is still preserved.

If the transmitting and receiving mechanisms of a phototelegraphy system are not running in synchronism, distortion of the received picture is the result. If, throughout the transmission, the receiving mechanism is running consistently faster or slower than the transmitter, the received picture, instead of being square or rectangular in outline, will appear diamond-shaped, and all the details will be twisted.

If the receiving mechanism is running at times faster, and times slower, the distortion will be even worse, and may easily render the picture completely unrecognisable. A similar wavy effect could be produced on a wet photographic negative which has been immersed in warm water to soften the film, by running the finger up and down it in parallel lines.

In phototelegraphy, given synchronism between the transmitter and receiver, the two mechanisms can be stopped at the end of the transmission of a picture, and simultaneously restarted for the transmission of the next, thus ensuring that both mechanisms begin at the proper starting-point in step with each other.

In television, however, both mechanisms are running continuously, transmitting and receiving, one after another, between twelve and twenty complete pictures per second, according to the arrangement of the system in use. Under these conditions it is possible for both mechanisms to be running at exactly the same speed and still the image may be incorrectly received at the distant receiver, due to a fault known as phase displacement.

If there is a definite lack of synchronism between the transmitter and receiver, the effect at the receiver is such that the observer sees a series of images flying past the viewing aperture, or across the screen, at a rate which depends upon the difference in speed between the two mechanisms. The further off synchronism the receiver is the faster will the images fly past. From the direction in which they fly, up or down in the case of vertical scanning and to right or left in the case of horizontal scanning, it is possible, knowing the direction of rotation of the disc, to tell whether the receiver scanning disc is running faster or slower than the transmitter disc.

If the receiver disc is in synchronism with the transmitter disc, but suffering from phase displacement, the received image will be stationary, but split. In the case of vertical scanning the effect is as if a man's face, instead of being visible squarely in the centre of the receiving screen, were displaced to right or left, so that his face appeared to be cut off vertically, say, by

FIG. 67.—The effect of phase displacement, when vertical scanning is in use.

FIG. 68.—The effect of phase displacement when horizontal scanning is in use.

the nose. On the other side of the screen the other half of his face would then be visible, also cut off by the nose. In the centre of the screen his right and left ears would be almost touching each other, as illustrated in Fig. 67.

In the case of horizontal scanning the effect is as shown in Fig. 68, which also illustrates a defect familiar to frequenters of cinemas, where the cause of the trouble is that the film is incorrectly "framed." A slight adjustment of the television receiver or the cinema projector rectifies the trouble in either case.

In phototelegraphy a similar effect would be obtained if, on starting the transmitting cylinder at the beginning of the picture, the receiving stylus were set, not at the commencing

METHODS OF SYNCHRONISM

end of the cylinder, but somewhere in the middle. If, when the recording stylus reached the end of the cylinder it were to be lifted and set at the other end of it, the correct starting-point, the result would be similar to that shown in Fig. 67.

This difficulty cannot arise in phototelegraphy, except through carelessness, for the correct starting-point is arranged first by hand before starting up. In television, with continuously running mechanisms handling anything up to twenty complete pictures per second, it is not possible to arrange the starting-point correctly by hand; some electrical or mechanical means of doing the job must be found. This is what was meant when it was stated that, in television, synchronism must be carried out, in a sense, to two dimensions.

So far we have used the popular general term " synchronism " to describe the requirements of television, and used the term very loosely. It is necessary now to define the requirements more accurately, and learn what the term synchronism really means in relation to television.

When two machines are running at exactly the same speed, the proper technical description is that they are running in *isochronism*. But although the machines may be in isochronism they may still be out of step, just as two men's legs may be out of step although both pairs of legs are moving at exactly the same speed and the feet of both men strike the ground at the same instant. A similar case is that of two clocks which are both keeping perfect time, although the hands of one might point to 2.30 and the hands of the other to three o'clock. Isochronism has been achieved in both cases, but for *synchronism* to be achieved the two men would have to be what the army calls " in step " and the hands of both clocks would have to indicate exactly the same hour.

In phototelegraphy it is only necessary to arrange automatic mechanism to preserve isochronism—hence the loose use of the word synchronism in this connection. Synchronism in phototelegraphy is actually achieved by hand by placing the recording head on the correct starting-point.

When the first efforts were made to achieve television, attempts were made to obtain isochronism by means of the methods used in phototelegraphy. Such methods, however, do not lend themselves to television, for, as already stated, they are either too complicated or insufficiently accurate, or both.

By using synchronous motors, however, perfect isochronism can readily be obtained, and the mechanical and electrical arrangements involved are not nearly so complicated. It was with the aid of such motors that the first successful demonstration of the transmission of silhouettes by wireless was given by Baird in April, 1925. For this demonstration he used two communication channels, one for the television signals and one for the isochronising impulses.

At first glance, it might be supposed that isochronism between two television mechanisms could be obtained by using two exactly similar motors controlled by rheostats and run at exactly the same speed, as indicated by some form of revolution counter. This cannot be done, however, for electric motors other than synchronous A.C. motors continually vary slightly in speed, due to small variations in the supply current and other reasons. These variations occur too rapidly to be adequately corrected by a hand-controlled rheostat. This habit of variation is known as "hunting," and before television can successfully be achieved, the hunting propensities of at least one of the machines must be brought under exact control.

The main driving motor at the transmitter has this usual tendency to hunt; and if the system of synchronism employed involves the use of an A.C. generator coupled to its shaft to produce the isochronising impulses, it may be allowed to hunt unchecked, for the periodicity of the generated A.C. varies in exact accordance with its speed wanderings. It is usual in modern equipment, however, as will be explained in detail in later chapters, to prevent the transmitter motor from hunting also.

On the other hand, the main driving motor at the receiver is not allowed to hunt independently. Its speed is under the iron control of the synchronous motor coupled to it ; and as the speed of the latter varies precisely in sympathy with the periodicity changes of the distant A.C. generator, it follows that the main driving motor must at all times be revolving at exactly the same speed as the main transmitter motor. Therefore, isochronism is achieved.

There remains now the question of synchronism. That is to say, although we now have the two machines running at exactly the same speed, we have, as yet, no means for adjusting the mechanisms so that they run *in phase* with each other.

As already stated, a difference in phase results in a shift of

METHODS OF SYNCHRONISM

the received image as a whole, and this image shift is very easily rectified by the simple expedient of rotating the receiver driving mechanism as a whole about its spindle until the picture comes into view in its proper place.

This operation is very similar to the method of adjusting the spark gap of an old-fashioned synchronous spark wireless telegraph transmitter. In the latter case, since the revolving electrodes were mechanically coupled to the shaft of the A.C.

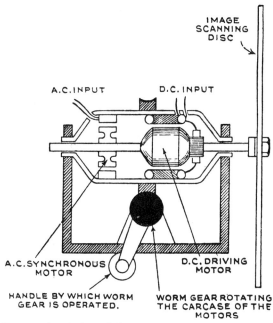

Fig. 69.—Cross-sectional view of the arrangement of a receiver main driving motor and synchronous A.C. control motor.

generator, the two were, of necessity, in isochronism at all times. The spark, however, might easily be out of phase, as indicated by an apparent violent oscillation back and forth of the revolving electrodes, as viewed end on. To bring the spark into phase, the fixed electrodes were simply rotated round the circumference of the revolving electrode disc until synchronism was achieved, as indicated by the disc appearing to stand still (stroboscopic effect).

In Fig. 69 is given a cross-sectional view of a television

receiver driving mechanism. At the extreme right-hand end of the shaft is the scanning disc. Farther to the left, within the carcase, is the main driving motor, which may be supplied with either D.C. or A.C., whichever is available. To the left of that is the synchronous motor which controls the speed of rotation of the main motor, giving isochronism.

The carcase of these motors is mounted on bearings, so that it can be rotated bodily by means of a handle operating through a worm gear. This feature is more clearly shown in Fig. 70.

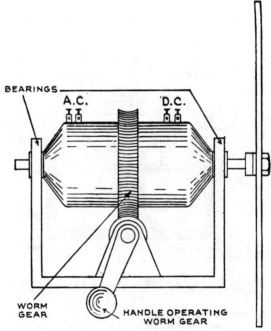

FIG. 70.—An external view of the driving and synchronising motors.

It will be seen that this mechanism has the merit of extreme simplicity, and it works extremely well in practice. It was first used by Baird in this country, and later by the American Telephone and Telegraph Co. in their 1927 and subsequent demonstrations.

In practice, after roughly running up to speed both transmitting and receiving mechanisms, and achieving isochronism by virtue of the control exercised by the synchronous motor at

METHODS OF SYNCHRONISM 119

the receiver, all the operator has to do to achieve synchronism is to watch the received image on the screen, and turn the worm handle shown in Figs. 69 and 70 until the picture appears on the screen in its proper place ; or in cinematograph parlance, until it is properly " framed."

There remains the question of the transmission to the receiver of the A.C. isochronising current generated at the transmitter. It is, of course, impossible at the present time to transmit power by wireless, or over a telephone line. Therefore, some means must be provided whereby the A.C. can be caused to influence the receiver. In the case of line transmission a very weak alternating current is sent over the line. In the case of wireless transmission the A.C. is caused to modulate the carrier wave of a wireless transmitter, as indicated in Fig. 55. At the receiving end the isochronising note, after amplification by means of valve amplifiers to build up the feeble A.C. to a sufficiently high-power level, is fed to the synchronous motor.

In order to bring the receiver driving motor into isochronism with the distant transmitter driving motor, Baird's method with his early apparatus was as follows.

First of all the receiver driving motor was run up to speed, under the rough control of a rheostat. The synchronous motor, of necessity, came up to speed as well, although as yet unsupplied with current. The input to the synchronous motor was controlled by a double pole switch, which connected it to the output of the isochronising impulse amplifier. Across the contacts of the switch were connected two little lamps.

As the synchronous motor came up to speed, the lamps flickered, the flickering becoming less and less (i.e. diminishing in frequency) as the speed of the synchronising motor (and, of course, the main driving motor) approached that of the generator at the transmitter.

When the speeds became exactly isochronous the flickering ceased and the lamps went out entirely. At that moment the switch was closed and the current from the amplifier fed to the synchronous motor. This current, if adequately amplified, was sufficient to prevent the synchronous motor creeping out of phase, and exerted sufficient force to prevent the receiver main driving motor from hunting independently of the transmitting motor.

It might be thought that a considerable amount of amplification would be necessary in order to obtain sufficient power to "hold" the mechanism at the correct speed. However, where well-balanced motors, equipped with ball bearings, are employed, only a relatively small A.C. current is necessary to keep them in step.

During the course of his original experiments, Baird used an isochronising frequency of 500 cycles. Generally speaking, the higher the frequency used, within limits, the more accurate is the degree of control. The American Telephone and Telegraph Co. used a frequency of 2000 cycles.

There remains to be discussed a more simple but not quite so effective a method of achieving isochronism, which has been widely used by American experimenters.

As already explained, the speed of synchronous motors depends entirely upon the number of poles with which it is equipped, and the frequency of the A.C. supply. Thus, two similar synchronous motors, fed from the same A.C. supply, will run at exactly the same speed. In America, electrical energy is supplied over wide areas at 110 volts, 60 cycles, so that providing the possessor of a television receiver takes his current supply from the same source as that which supplies the transmitter he can achieve a moderately successful degree of isochronism by using a similar type of synchronous motor to that in use at the transmitter, and controlling it by means of a fine rheostat. However, sudden heavy loads applied to the mains in the vicinity of the receiver cause temporary variations in speed which have to be corrected by means of the fine rheostat, and the operation of this control approximates somewhat to a juggling feat, although it is surprising how steady the image can be held with some practice.

Apart from the isochronism problem, however, there is at the low frequency of 60 cycles a phenomenon known as phase swinging which causes even synchronous motors to vary in speed very slightly from instant to instant. The effect of this is to cause the received image to swing or hunt slightly but rapidly in and out of its "frame." This effect is very distressing and tiresome to the eyes of the observer.

On the principles outlined in this chapter all synchronising methods at present in use are based. The actual details of the different methods will be given in later chapters.

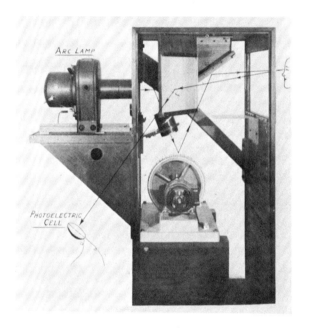

[*Courtesy: Telefunken Company.*

One form of Karolus-Telefunken television transmitter, using a Weiller mirror wheel for scanning purposes. Compare this with the diagram in Fig. 93.

[*To face page* 120.

CHAPTER X.

IMAGE STRUCTURE.

Methods of Signalling. Structure of Half-Tone Reproductions. The Dot Theory of Television. " Picture Points." Aperture Distortion. Picture Ratio. Horizontal v. Vertical Scanning.

ONE of the greatest controversies in connection with the earliest development of television arose over the composition or structure of the received image, it being argued by some who had never seen a television image, and who could not credit the possibility of even a crude form of television, that in order to produce a recognisable image it would be necessary to transmit such a stupendous number of electrical impulses per second that no wire or wireless circuit could possibly carry them. It is proposed, therefore, to devote this chapter to an examination of image structure, and some of the problems associated therewith.

The erroneous theory mentioned above has long since been exploded by countless demonstrations, both in this country and abroad, of limited but recognisable television images. The theory was a legacy from the older art of phototelegraphy as transmitted by what is known as the interruption method of signalling. In this method the electrical impulses representing the picture are sent as discontinuous signals, the carrier current (in the case of wire transmission) or the wireless wave (in the case of wireless transmission) being cut off entirely between one impulse and the next.

This method is not used to-day in high-quality phototelegraphy systems, which now use what is called the modulation system of signalling. In this system the electrical impulses are caused to modulate or vary the intensity of an otherwise steadily flowing carrier current, or the amplitude of a wireless carrier wave. Ordinary sound broadcasting is a typical and universally familiar example of the modulation method of signalling.

In passing, it should perhaps be explained that the difference in principle between the two systems is not so great as would at first appear. In Fig. 71 we have a steady unmodulated, or uninterrupted current represented by the straight line a. This current may be a D.C. line carrier current, or a high-frequency wireless carrier wave. If this current is interrupted to form telegraphic signals, the resultant current curve may be as shown in Fig. 72.

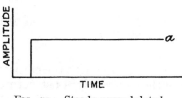

FIG. 71.—Steady unmodulated carrier current.

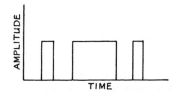

FIG. 72.—Carrier current interrupted to telegraph in Morse code the letter R.

If signalling is to be carried out by modulation, however, the steady current shown in Fig. 71 will deviate or vary in amplitude gradually, in exact accordance with varying degrees of modulation, as shown in Fig. 73. The lowest or highest point of the curve, representing the greatest change in amplitude, is usually referred to as the maximum percentage of modulation;

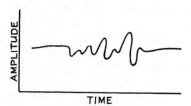

FIG. 73.—Modulation method of signalling.

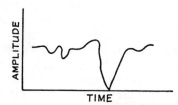

FIG. 74.—Illustrating momentary 100 per cent. modulation.

i.e. if the amplitude of the steady current is increased or reduced by half, the maximum degree of modulation is 50 per cent.

If the maximum degree of modulation is increased for an instant to 100 per cent., as shown in Fig. 74, it means that the amplitude of the carrier current, or wave, is for an instant reduced to zero. That is to say, nothing is transmitted or radiated.

IMAGE STRUCTURE

If the maximum degree of modulation is allowed to exceed 100 per cent. there is a definite interruption of the signal, as shown in Fig. 75, where maximum modulation is shown as having been arrived at gradually, in steps. If modulation in excess of 100 per cent. is arrived at *instantly*, we immediately revert back to the interruption method of signalling, as illustrated in Fig. 72.

FIG. 75.—When modulation exceeds 100 per cent., a reversion to the interruption method of signalling takes place.

From the above explanation it is clear that the interruption method of signalling permits only of the transmission of signals of variable duration ; the intensity of all signals is constant. The received picture, therefore, consists of a pattern of lines of irregular length with white spaces of varying extent between them. No fine half-tones are possible. The number of impulses which can be transmitted per second is strictly limited, which means that it takes a considerable period of time (running into many minutes) to send a single picture.

The modulation method allows the transmission not only of impulses of variable length, but also of variable intensity, with the result that the most delicate half-tones can be transmitted and received. The received picture is recorded photographically, and in the best systems in use to-day one requires very good eyesight indeed to tell the difference between a photograph received telegraphically and the original. The number of impulses transmitted per second can be, and is, very much higher, with the result that a picture can be transmitted in a matter of seconds if the distance to be covered is not too great. As we shall see later there is, at present, a definite limit to the number of impulses which can be sent per second, even by the modulation method.

Television makes use of the modulation method of signalling, and the received image is therefore not a black and white picture made up of short lines or dots, but an image made up of softly graduated half-tones. Because this was stated at the time when television was first successfully demonstrated, critics referred back to phototelegraphy as conducted by the interruption method of signalling by the earliest workers in that art. In order to get half-tone effects, it was argued, television workers would have to divide up the image into dots of varying sizes, as is done in the case of a newspaper illustration. Each dot would have to be transformed into an electrical impulse, and the total number of impulses which it would be necessary to transmit per second in order to get even a small and coarse-grained image would be so enormous as to be beyond the carrying capacity of any signalling circuit, either wire or wireless.

A newspaper illustration was quoted as an example of the coarsest grained picture which would be acceptable. Actually, all half-tone illustrations (reproductions of photographs) are composed of dots, which are the result of the process through which the blocks go from which the reproductions are printed. This process is as follows.

The photographic print which is to be reproduced is first of all photographed by means of a special camera, using a special type of photographic plate. Inside this camera, between the lens and the plate, and almost touching the emulsion of the plate, there is placed a screen composed of two sheets of glass, both ruled with very fine lines. These sheets are so arranged that the ruled lines are at right angles to each other. This gives the appearance of a number of tiny squares, or dots. On the number of lines per inch depends the "screen" of the reproduction, i.e. the degree of coarseness or fineness of the "grain." The screen chosen depends upon the quality of the paper upon which the finished block is to be printed. For newspaper work, on cheap, coarse-fibred paper, the screen usually employed is 55, which means that the reproduction is built up of 55 dots per linear inch, or 3025 per square inch. Very fine art paper, such as is used for the half-tone illustrations of this book, allows a much finer screen to be used. To be exact, the screen used is 133. In Fig. 76, which is given to illustrate the subject in clearer detail, three views of portions of a man's face are given, one with a standard 55 screen, and two with specially made screens ruled with 16 and 8 lines per inch respectively. The reader will have to hold the book some distance away before the dots merge sufficiently to make the pictures intelligible.

When the plate is taken from the camera and developed, a negative is obtained which is covered with tiny dots of different sizes. In black parts of the picture the dots merge into one another, and give a solid black effect. In the shaded, or half-tone

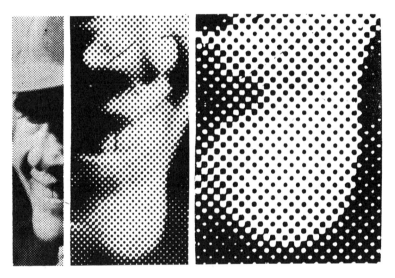

FIG. 76.—Three half-tone blocks having screens 55, 16, and 8 respectively. They are all parts of the same photograph and serve to illustrate how picture structure is built up.

[*To face page* 124.

IMAGE STRUCTURE

parts of the picture, the dots are smaller and fairly widely separated; in the bright highlights the dots are extremely minute in size and widely spaced.

A sheet of specially polished copper (for the finer screens, over 85) or zinc (for the coarser screens) is now selected and sensitised, and a contact print made from the negative, as in ordinary photography. The sheet of copper or zinc is then "developed," and placed in an acid bath and "etched." During the etching process the metal between dots is eaten away to a certain predetermined depth, leaving untouched those parts of the surface which are preserved from acid action by the action of light during the printing process. These parts project from the etched surface in the form of little pinnacles with round flat tops.

When the etching process is finished, the sheet of copper or zinc is nailed to a block of type-high wood, ready for printing. When printer's ink is rolled over the surface of the metal, and a print or "pull" is taken on a sheet of paper, the tiny flat pinnacles deposit their ink on it, leaving white paper between them, as illustrated in Fig. 76. If the finer screen illustrations in this book are examined with the aid of a powerful magnifying glass, it will be seen that their structure is built up in exactly the same way.

This digression from our subject is not quite so irrelevant as might at first appear, because it enables us to appreciate two things; firstly, the more detail we want in the reproduced picture the larger the number of picture elements (dots) we must employ per unit area; secondly, the maximum number of picture elements we can reproduce cleanly per unit area, without blurring, depends upon the quality, or shall we say the "performance factor," of the medium employed, which in this case is the kind of paper used for printing.

If we consider a television image only 2 inches square, built up of dots after the fashion of a newspaper reproduction of screen 55, the total number of dots will be $55^2 \times 2^2 = 12,100$; and if, in order to give the illusion of natural movement, this image is transmitted sixteen times per second (formerly standard silent cinema film speed) the total number of electrical impulses (representing the dots) which we must transmit per second, reaches the stupendous figure of 193,600. Such a speed of signalling is (at present) quite impracticable, except on ultra-short waves.

The literal accuracy of this theory of television, which became known as the dot theory, was assailed and finally exploded by Dr. J. Robinson.[1] A television image, as at present built up, is made up, not of dots, but of strips, the width of each strip being equivalent to the diameter of the holes in the scanning disc. If the scanning disc has been accurately made the strips fit exactly edge to edge, without either overlapping or underlapping.

An image made up of dots would necessitate the use of the interruption method of signalling instead of the modulation method. Actually, each strip is scanned continuously, and not in discrete patches, sections, or dots; the variation in light intensity is gradual over the length of the strip, the degree of variation and the number of changes depending upon the contrast, or light and shade details of the scene being transmitted. The varying or undulating current output of the photoelectric cell, or cells, produced by the continuously varying light impinging on them, is caused, after amplification, to modulate the carrier wave, producing continual variations in amplitude such as are illustrated in Fig. 73.

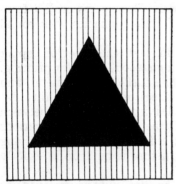

Fig. 77.—A simple figure used to disprove the dot theory of television image structure.

For example, let us consider Fig. 77, a simple case in which the image of a black triangle is being transmitted. In the Baird system, scanning takes place from right to left, commencing at the lower right-hand corner and moving upwards. The first four strips contain no image, so that four long electrical impulses only are necessary to reproduce four plain white strips. In the fifth strip there will be a sudden drop in the amplitude or intensity of the image current for a brief interval, which will serve to reproduce the lower right-hand tip of the triangle. This sudden drop will be signalled to the receiver.

---

[1] *Television*, Vol. I., Nos. 9, 10, and 11.

[*Courtesy*: *Telefunken Company.*

Another form of Karolus-Telefunken television transmitter, giving a close-up view of the mirror wheel and synchronising arrangements. Under the wheel is the casing of the synchronous motor employed to keep the speed of the main driving motor (below synchronous motor) constant.

IMAGE STRUCTURE

For the sixth and subsequent strips, up to the fifteenth, the duration of the current drop will be progressively longer.

The important point is that, to represent any given strip between the fourth and the twenty-sixth, only two changes in the value of the output current of the photoelectric cells are necessary, one a sudden decrease as the scanning beam at the transmitter encounters the edge of the black triangle, and the other a sudden increase in current as the scanning beam leaves the other edge.

Actually, the total number of current changes required to transmit the simple scene depicted in Fig. 77 (which shows a square picture made up of thirty strips, and there is a short interval between each strip) is 74. If sixteen images of Fig. 77 were to be transmitted per second, the total number of impulses sent in one second would be 1184; but as the output current of the photoelectric cells, after amplification, is alternating in character (the D.C. component is suppressed), the actual frequency to be transmitted, neglecting possible complications due to harmonics, would be half that figure, or 592 cycles.

The reasoning used in this example can be applied with equal accuracy to the most complicated scenes. The more complicated the scene, of course, the higher the rate of change, or transmission frequency, if all the detail is to be faithfully reproduced, just as in the case of process blocks for printing, a finer screen (more dots per inch) and better quality paper are required to bring out fine detail. In the case of television, the paper must be replaced by high quality terminal apparatus at the receiver, and a high quality transmission channel capable of dealing satisfactorily with the increased number of impulses which have to be transmitted per second. But in every scene there are more or less extensive areas which have the same density, or tone value, so that over several arbitrary linear units of each scanning strip there will be but one current impulse, and not a number of impulses equivalent to the number of arbitrary units into which the strip may be divided for mathematical or other convenience.

To be more explicit, there are many technicians, especially in Germany, who consider each strip, in a square image made up of thirty strips, to be divided into thirty units of " picture points," as they are called in Germany. The total number of picture points in Fig. 77, by this arbitrary method of calculation, would thus be $30 \times 30 = 900$, instead of 74 as we found above.

At a transmission speed of 16 images per second the total number of impulses would thus be $900 \times 16 = 14{,}400$, and the A.C. frequency would be 7200 cycles instead of 592.

The picture point theory is as obviously wrong as the dot theory, but its supporters, although some of them admit the weakness of the idea, urge in defence of it that some form of quantitative measure is necessary, even if it is only arbitrary, in order to classify different systems and gauge, if only roughly, their relative efficiencies in terms of the amount of detail obtainable in the image. But it does not serve even that purpose, for A can easily build a system having more picture points than B's system, but showing an identical degree of detail, by simply lengthening the image along the direction of the strips, keeping the number of strips the same.

In America it is customary to indicate the degree of detail in the image by quoting the number of image strips (i.e. holes in the scanning disc) which are employed, which is a sufficiently accurate method of qualitative judging so long as scenes no larger than the present head and shoulders' view of human beings can be transmitted. Since the size of the scene, or field of view is constant, the larger the number of strips into which it is divided, the greater will be the detail reproduced at the receiving end.

Another purpose which the picture point designation is intended to fulfil is to indicate the frequency of the image signals, but as shown above the result obtained by working on this hypothesis is not strictly correct. The uncertainty and confusion which exists on this point is indicative of the embryonic state in which television still remains. Those engaged upon research into the many problems involved are still probing more or less uncertainly in the dark, looking for a leading light which will guide them towards intelligent solution. This state of affairs is not exclusively exceptional to television; every new science has had to go through the same difficulties during its early stages.

Before leaving the question of image detail, or structure, there is another point which must be considered, and that is the phenomenon known as aperture distortion.

If the holes, or apertures in the transmitting and receiving scanning discs could be made infinitely small in diameter, the amount of detail obtainable in the received image would be infinitely great. But an infinitely small aperture is impossible

IMAGE STRUCTURE

to achieve in practice; it must have finite dimensions. As a result of this limitation a form of distortion of the image occurs which is due to the passage of the scanning light-spot (of finite size because of the finite size of the disc aperture which creates it) from a bright part of the scene to a dark part or *vice versa*. Instead of a change of photoelectric current from maximum to minimum taking place instantaneously, the change takes place gradually, beginning when the advancing edge of the scanning spot leaves the bright part of the scene and commences to enter upon the dark part. As the spot continues to travel, so the photoelectric current falls gradually, until the whole of the spot has left the bright part. The intensity of light reflected back to the photoelectric cell from the transition part of the scene is, as it were, gradually eclipsed.

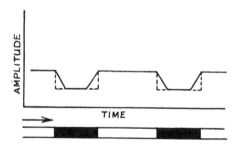

FIG. 78.—Illustrating the effect of aperture distortion on the photoelectric current.

The effect is illustrated diagrammatically in Fig. 78, where an image strip composed of alternate white and black areas is shown being scanned in the direction of the arrow. The resultant curve, corresponding to the output current of the photoelectric cell, is shown above the strip. If the scanning aperture were infinitely small, the curve would take the ideal form, where the tone value changes, shown by the dotted lines; that is to say, the current changes would be abrupt and instantaneous. As it is, with an aperture of finite dimensions, the decreases and increases of current are gradual, as shown by the slanting full lines of the curve.

The effect of aperture distortion upon a received image is that sharply defined lines of demarcation between bright and dark parts of the original scene are blurred or softened in the received television image. This is illustrated (in an exaggerated form) in Fig. 79, which shows five image strips. In the upper part of the drawing is a black diagonal bar as seen at the transmitter. Below is the same bar as reproduced at the receiver. The edge line now takes the form of zig-zag steps, but will be somewhat less harsh in outline than it is possible to show in a sketch such as Fig. 79.

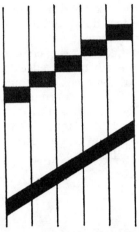

FIG. 79.—Illustrating the effect of aperture distortion on a received image.

Aperture distortion is not confined to television. It can also be detected in the very highest quality phototelegraphy systems in use to-day, but as the entire process is very much slower, and very fine strips are employed, it almost requires a powerful magnifying glass to reveal the defect. The effects of aperture distortion can, however, be very largely compensated for (but not entirely eliminated) by the proper use of correction networks, or " equalisers," in the photoelectric cell amplifier circuits.

It only remains now to point out that, as an aperture of infinitely small dimensions is the unattainable ideal for television, the best results, from the point of view of clear definition, will be obtained by the use of as small an aperture as possible. But if the aperture is made too small, quite apart from the mechanical difficulties of drilling very small apertures accurately, a point will be reached when the loss of light, both at the transmitting and receiving ends, will become a serious factor. This is one of the limitations of the present Nipkow disc system, and until a different system is devised, we must be content with a compromise. Even then, any system which builds up an image from either strips or dots will suffer from the defect to an extent which will be dependent upon the degree of fineness of analysis and synthesis of the image.

However, it must not be thought that the Nipkow disc system is incapable of producing acceptable results; on the contrary, modern electrical and mechanical engineering ingenuity is capable of producing discs and associated equipment which will give admirable results, although the field of view which

IMAGE STRUCTURE

can be reproduced with such good results is restricted to the human head and shoulders.

An idea of the amount of detail which can be reproduced may be obtained from the fact that during a recent test in the Bell Telephone Laboratories, using a 72-hole disc (to be described in a later chapter), lip readers who watched a silent image were able to understand what was said by the person seated before the distant transmitter. The detail was sufficiently good to enable them to follow the slightest movements of the lips and interpret the words of the unheard speaker.

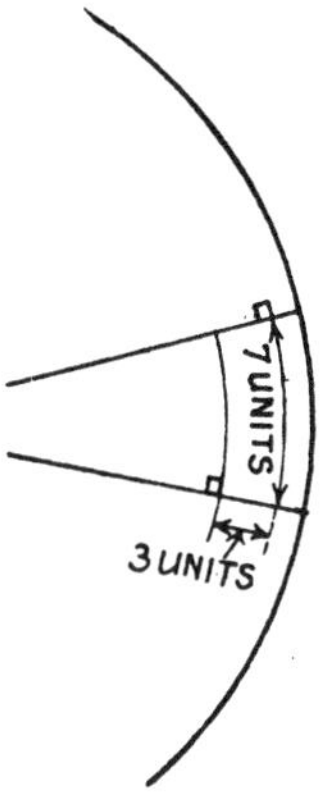

Another factor of image structure which remains to be considered is the picture ratio, which is simply the ratio between the height of the picture and its width. In the case of the Baird system this ratio is 7 to 3, which means that the circumferential distance

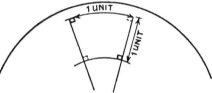

FIG. 80. — Illustrating the Baird picture ratio, which is 7 to 3; the circumferential distance between one hole of the scanning disc and the next is 7 units, while the radial distance between the outermost and innermost holes of the spiral is 3 units. This gives a narrow oblong picture of which the two holes shown here form diagonally opposite corners. This diagram also illustrates vertical scanning.

FIG. 81.—Illustrating the 1 to 1 picture ratio common in America. The picture is as nearly square as the arrangement of holes on a circular disc will allow. Horizontal scanning is illustrated in this diagram.

between one hole of the scanning disc and the next is 7 linear units, while the radial distance between the outermost and the innermost holes of the spiral measures 3 linear units; in other words, the image is 3 units wide by 7 units high. The standard picture ratio laid down for use in Germany is 3 to 4, while most American experimenters use a 1 to 1 ratio; that is to say, they produce a square picture. Figs. 80 and 81 will serve to illustrate what is meant.

Besides using a picture ratio entirely different to anyone else, Baird is also practically the only worker who scans his images vertically; the others scan horizontally. In the case of vertical scanning by means of a disc, the received image appears at the side of the disc, and the " grain " or strip lines of the image structure show vertically. When horizontal scanning is employed by means of a disc, the received image is seen at the top of the disc, and the lines run horizontally. Figs. 80 and 81 illustrate both methods.

These differences of standards are naturally confusing to anyone approaching the subject for the first time. Undoubtedly a universal standard will ultimately be adopted; otherwise international television broadcasts will be impossible unless one has an interchangeable receiver—a cumbersome and unsatisfactory arrangement somewhat comparable with the present state of broadcasting in Europe, where one has, with many receivers, to change coils in order to change from medium to long waves, and *vice versa*.

The choices made by different workers have, up to the present, been largely haphazard, and for the benefit of students of the subject, the considerations leading to the choice of these various standards will be briefly outlined. The first man to use horizontal scanning appears to have been Jenkins, who used it in his first shadowgraph demonstration in 1925, and so far as the author has been able to ascertain by personal enquiries, the choice seems to have been purely haphazard, probably due to some feature of the construction of the apparatus which made it more convenient to use the top of the disc. The fashion having been started, others followed suit.

The Germans, surveying the chaos which existed at the time when they commenced to take an official interest in the subject, made the choice of their standards as the result of intelligent and typically methodical reasoning. The picture ratio which they finally adopted, 3 to 4, is the standard motion picture ratio, and was selected partly because it is the ratio of height to width to which everybody has become accustomed, and expects in any picture, still or moving, as witness the popularity of the $3\frac{1}{4}$ inches $\times$ $4\frac{1}{4}$ inches (quarter plate) camera ; and partly because the German plan is to broadcast cinematograph films in the first place, during the initial experimental broadcast transmissions.

Having thus settled the ratio, horizontal scanning follows almost as a matter of course, for in order to obtain the greatest detail the picture strips must run the long way of the picture ; to scan vertically would entail the employment of larger holes

[*Photos Courtesy: Telefunken Company.*

Two views of the Karolus-Telefunken large screen television receiver. At left of upper picture is the arc lamp and Kerr-cell Nicol-prism combination. Bottom picture shows the mirror wheel and its driving and synchronous motors, and the path of the light-spot. Compare with Fig. 94.

[*To face page* 132.

in the scanning disc, with consequent loss of detail. The alternative would be to keep the diameter of the holes the same, but increase their number, with consequent increase in the frequency band to be transmitted.

For the transmission of television only (and not cinematograph films) the limitations imposed at present by the available frequency band width make it impossible to transmit on the broadcast waveband, with a reasonable amount of detail, any scene much more extensive in area and detail than a head and shoulders view of a human being. Now, if the human face be considered for a moment, it will be realised that the most prominent characteristic lines which go to make up the features (hair over forehead, eyebrows, eyes, nostrils, mouth, chin line) run in a horizontal direction, and it is demonstrable fact that strongly marked lines can best be scanned and reproduced in a direction at right angles to them. This is because of the distortion created by the aperture effect. Under these circumstances vertical scanning, as adopted by Baird, is obviously best suited to present conditions.

As to the picture ratio under these circumstances, a few moments' study of a head and shoulders photograph of square dimensions will reveal that quite a lot of paper can be cut off both sides without in any way detracting from the likeness and pictorial value of the photograph (see Fig. 76). A ratio of height to width of 3 to 1, or even 4 to 1 will still preserve all the relevant details. Baird has finally settled down to using a ratio of 2·3 to 1 (7 to 3) for the above reasons, and for reasons of economy of frequencies and simplicity of apparatus, but a narrow oblong picture such as is given by a 7 to 3 ratio does not permit of very much lateral movement on the part of the subject being televised—a disadvantage which handicaps an artiste who is trying, by visual means, to augment his vocal effort to broadcast an item which will entertain the recipient.

In contrast to the above reasoned arguments which are advanced to-day in favour of vertical scanning and a narrow oblong picture, it is of interest to note that these practices originated during the course of Baird's early experiments, using lens discs, because he was anxious to get the most he could out of a disc containing as few lenses as possible! Although it is true that Baird is a Scot, it must be said in his defence that at that time he was living in poverty and lenses were expensive. Also more lenses on a disc meant that more power was required to drive the disc, greater centrifugal force resulted during rotation, and as the discs were made of heavy cardboard the chances of a disc flying to bits were greatly increased.

There is another highly interesting point which has a bearing on the Horizontal $v.$ Vertical scanning controversy.

If the electric motor which drives the television receiver has the slightest tendency at all to hunt, the movement imparted to the image, in the case of horizontal scanning, is to and fro horizontally; in the case of vertical scanning the movement is vertical. Now, it is a curious fact that, through long usage, we can more readily detect horizontal movement. Most movements in our everyday life take place in a horizontal plane, and so we have become accustomed to detecting them and moving our eyes from side to side rapidly. We are not accustomed to such rapid and continual movement in a vertical plane, in spite of the increasing numbers of aeroplanes which tend to attract our attention to the skies. On the contrary, we subconsciously oppose small vertical movements because, when we walk, our heads move slightly up and down; our eyes have become so accommodated that we do not notice the resultant slight vertical movement. As a result, a given amount of hunting in a television image becomes far more noticeable and distressing in the case of horizontal scanning than it does where vertical scanning is employed. By making comparative tests with different systems, the author has proved conclusively that this is no imaginary phenomenon, but a very real one.

CHAPTER XI.

TRANSMISSION CHANNELS.

Frequency Limitations of Existing Channels. Use of Submarine Cables and Long Wireless Waves Impossible at Present. Distances Covered to Date by Television. Difficulties on Short Waves. Medium Waves Best. The Sideband Theory. All Wavebands Overcrowded. The Stenode Radiostat Explained. Bell System Experiments with Multiplex Transmission, and Conclusions. Stenode Radiostat Principle Promises Vast Extension of Transmission Channels and Band Widths. Wire Channels Best at Present.

REFERENCE has been made to frequency band width limitations in connection with the broadcasting of television. Before proceeding to a discussion of the problems which lie in this direction, it might perhaps be well to point out here, clearly and very definitely, that the obstacles in the way of the further development of television so as to enable larger scenes to be transmitted and received with greater detail on larger screens, are divided almost equally between the terminal equipment and channels of communication. The limitation exists whether the link be a wire or wireless one.

By means of existing television terminal equipment it is already possible to deal with a scene larger than a head and shoulders view of a human being, and in greater detail than is at present being broadcast in this and in other countries. Furthermore, the ultimate limit of development of existing equipment has not yet been reached. Obviously, therefore, we must examine the system or medium of transmission and see wherein it fails to meet the requirements of television.

As already explained, modern systems of phototelegraphy, and also television, make use of the modulation method of signalling, whereby the picture signal is superimposed upon a carrier current or carrier wave, just as are speech currents in telephony or sound broadcasting. During the process of scanning a single strip of a scene to be transmitted by television, the light reflected back from the scene varies in intensity over the whole length of the strip, in exact accordance with the light and shade of the scene. This reflected light, falling on the photoelectric cell, gives rise to a unidirectional current of continually varying amplitude. The frequency of the amplitude changes also varies continually. But the electrical output of a photoelectric cell, or even a battery of cells connected in parallel, is so extremely feeble that it has to be amplified very considerably. One result of amplification is that the unidirectional current of varying intensity originally delivered by the cell is transformed into an alternating current of varying amplitude and frequency. In this respect the output is similar to the amplified output of a microphone.

The amplitude depends upon the degree or intensity of light and shade in the original scene, and the frequency depends upon the number of changes from light to dark, or intermediately, which have to be recorded per second. On amplitude depends *contrast* in the received image; on frequency depends *detail*.

For example, if a blank white screen is held in front of a transmitter, and a square image of it is scanned with thirty lines sixteen times per second, the amplitude of the image signal will be constant, and its frequency will be constant at $30 \times 16 = 480$ cycles per second. In the case illustrated in Fig. 77 the amplitude varies in accordance with the design which has to be scanned, and the frequency is 592 cycles. As the design or scene becomes more complicated, and especially when half-tone effects are introduced, the variations in amplitude become more complicated, and the frequency of these changes increases. For a given frequency limit we can transmit either a small scene with a wealth of detail, or a large scene with scarcely any detail.

The inevitable compromise is the largest possible scene which can be transmitted with just sufficient detail to be recognisable, and that, under present conditions, is the human head and shoulders. It is a fact that if something of equal dimensions, but with which we are not nearly so familiar, were placed in front of most of the present-day television transmitters, the detail given at the receiver would be insufficient for an observer to recognise any unfamiliar object.

From the foregoing it will be seen what a vital part is played by the frequency of the television signal. It is now time to consider what are the factors which at present prohibit an unlimited increase in frequency, and therefore detail.

If transmission is to be carried out by wire, then, under existing conditions, the frequency is largely limited by the distance which has to be covered. It must be remembered that the more delicate degrees of light and shade (half-tone effects) in a scene to be transmitted by television produce correspondingly delicate degrees of modulation of the carrier current, and when transmission is carried out over a wire or cable these finer distinctions of current values tend to become smoothed out and lost. This smoothing effect is caused by the distributed inductance and capacity of the wire or cable circuit.

This has long been one of the greatest difficulties which the cable companies have had to contend with in their efforts to speed up signalling over submarine cables. The fastest existing cable, known as a permalloy loaded cable, has a maximum operational speed of about 500 words per minute for telegraphy. At this speed, only about 125 impulses per second are transmitted. For just moderately intelligible speech a channel of communication is necessary which will carry a frequency of not less than 1500 cycles.

Up to the present the impossibility of designing a submarine cable to carry such a frequency has made it impossible to establish a transatlantic telephone service except by wireless. However, a new magnetic material, known as perminvar, has just been developed by the Bell Telephone Laboratories which, when used to load a submarine cable, will make transatlantic submarine telephony possible, and arrangements are now being made to make and lay such a cable. But to transmit across the Atlantic, by cable, a television image such as is now being broadcast from London would entail the provision of a cable capable of conveying, undistorted, a frequency band width of 9000 cycles.

Over land wires, such as high-grade telephone circuits, much wider frequency bands can be transmitted, and more attention has been devoted to extending the frequency characteristics, largely through the necessity for the use of such circuits for the relaying of broadcasting entertainment, which embraces the same frequency band. In America, as we shall see in a later chapter, the Bell Telephone Laboratories have developed telephone circuits capable of carrying television frequencies as high as 40,000 cycles. On land, apparatus can be inserted in the circuit at regular and frequent intervals to correct the attenuation and distortion which would otherwise result The difficulties in the way of treating submarine cables in a similar manner are obvious.

As a matter of historical interest, the longest distance over which television has been transmitted, through the medium of a telephone circuit, is from London to Glasgow, a distance of about 435 miles. This was accomplished by Baird in May, 1927.

When television is to be transmitted by wireless, there are several factors which have to be considered. The first question to be settled is that of the wavelength to be employed. For long-distance work, using a reasonable amount of power, short waves would appear to be indicated. But short waves, as every wireless enthusiast well knows, suffer considerably from fading, which causes loss of detail in the received image; the image literally fades, or even disappears altogether, leaving a blank screen. There is also the phenomenon known as "skip distance," and this, together with fading, precludes the use of short waves for the satisfactory broadcasting of either speech or television. For fixed point to point work, using beam wireless, these difficulties can very largely be overcome.

For all-round even reception, up to distances of several thousand miles, very long waves must still be employed for consistently reliable results. But such waves are only suitable for telegraphic communication, because the frequency which can be superimposed on the carrier wave is limited, partly by the transmitting aerial, and partly by the frequency of the carrier wave itself. It is partly because of this limitation that the speed of wireless telegraph signalling is not so high on very long waves as it is on short waves. Modern high-power, sharply-tuned, long-wave transmitting aerials, when an impulse is delivered to them from the transmitter, take an appreciable time to charge up, owing to their large distributed capacity. If the speed of transmission (or frequency of modulation) is increased beyond a certain limit, depending on the wavelength and the characteristics of the aerial, the latter will not have time to become fully charged, i.e. full aerial power will not be built up. This means that individual impulses will be attenuated, or weakened, before they even leave the transmitting aerial. If an attempt is made to modulate the carrier with too high a frequency, extending beyond the width of the resonance curve of the aerial, an attenuating or direct cut-off effect will take place, resulting in the suppression of the higher fre-

[*Courtesy: Telefunken Company.*

A combination receiver produced (but not marketed) by the Telefunken Company. The upper opening contains the television receiving screen. Under it is the loud speaker.

TRANSMISSION CHANNELS

quencies. This effect is similar to that which occurs in many badly designed broadcast receivers in use to-day, and which results in the poor quality reproduction of music.

Considering the other limitation to the use of very long waves, the frequency of the carrier wave itself, it is obvious that in order to transmit 10,000 signals, or impulses, per second, as required for crude head-and-shoulder television, we must use a carrier wave having a frequency of not less than 10,000 cycles (which corresponds to a wavelength of 30,000 metres), for we must have at least one wave per signal. Wireless signalling would not, in fact, be possible at the rate of one wave per signal; it is advisable to have a train of waves consisting of a considerable number of waves for each signal.

The position was very clearly expressed by Dr. J. Robinson in the November, 1929, issue of *Television*. He wrote as follows:

" For the present definition of the television image, corresponding to 2000 picture elements and dealing with about 20,000 picture elements per second" (transmission frequency 10,000 cycles per second), " when the wavelength is 300 metres there are trains of waves available of 100 waves each for each pair of picture elements, and this is a reasonable value." (The frequency of a 300-metre wave is 1,000,000, which, divided by 10,000, gives the figure 100.) " For a wavelength of 3000 metres the trains would consist of 10 waves, which is getting low; and for a wavelength of 30 metres the trains would contain 1000 waves. Thus it is obvious that the wavelength places a definite control on the definition of the television image, and as this definition is improved, say to 10,000 picture elements, it will be physically impossible to employ long waves."

The first spectacular effort to transmit television by wireless over a long distance was made by Baird on February 9th, 1928, when, using a wavelength of 40 metres, he succeeded in transmitting television images 3500 miles across the Atlantic to an amateur receiving station near New York. For the purpose of this experiment, Baird used a 1 Kw. wireless transmitter belonging to an amateur at Purley, near London.

On the night of March 5th-6th, 1928, the experiment was repeated, this time to the Cunard liner *Berengaria*, then 1500 miles out in mid-Atlantic. The television receiver used in New York was then on its way home, and passengers and officers of the liner had an opportunity to look in to the broadcast images of persons seated in London. Although both experiments were reported to have met with some measure of success, considerable difficulty was experienced due to fading and atmospherics. The present author had the pleasure of being present on both of these historic occasions at the London transmitting studio, and was one of those who sat before the transmitter.

In passing, it should be recorded that, despite the various disabilities of short waves, the American Government has set aside certain short-wave bands for the use of television experimenters, who are permitted to send out frequency bands up to 100 Kc. wide on the specified wavelengths. (See Appendix to Chapter XIV.)

Early in 1930 Dr. E. F. W. Alexanderson, of the General Electric Company, experimentally broadcast the image of a black rectangle on a white background, from a short-wave transmitter at Schenectady, N.Y. The image signals were picked up at Wellington, New Zealand, rebroadcast to Sydney, Australia, whence they were again rebroadcast on short waves and received at Dr. Alexanderson's laboratory at Schenectady. The total distance covered by the television signals was 20,000 miles, and it is remarkable that the received image was at moments a recognisable replica of the original rectangle. Other aspects of this experiment will be dealt with in a later chapter.

With long waves ruled out as physically impossible, and with short waves at present unsuitable for broadcast purposes owing to fading, skip distance, and echo effects, it is obvious that, for the time being, we must use medium waves; and if television is to be broadcast for entertainment purposes, the most suitable of the medium waves are those used for sound broadcasting, viz. 250 to 550 metres. But these wavelengths are already filled to overflowing with broadcasting stations, and it has been necessary, in Europe, to limit all stations by international agreement to the transmission of a waveband not more than 9000 cycles wide.

As many readers of this book may not be sufficiently acquainted with wireless technique to understand the meaning of the term "waveband," and the limitations imposed at present by what is known as the "sideband theory" of wireless transmission and reception, it will now be explained in detail with the aid of a concrete example.

A 300-metre wave has a frequency of 1,000,000 cycles. If this wave is modulated with a musical note, or frequency, of 4500 cycles, then, according to the sideband theory, a combination of frequencies is radiated or transmitted which amounts to 1,000,000 plus or minus 4500. That is to say, frequencies anywhere between 1,004,500 and 995,500 cycles will be radiated. The difference between these two figures is 9000 cycles, or 9 kilocycles, the present frequency separation between European broadcasting stations.

The resonance curve of an unmodulated carrier wave has an appearance similar to the curve shown in Fig. 82, but, according to the sideband theory, as soon as the wave is modulated the curve takes on the appearance shown in Fig. 83.

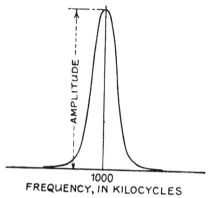

FIG. 82.—The resonance curve of the unmodulated carrier wave of a sharply tuned broadcast transmitter. Nearly all the radiated energy is concentrated on a single frequency, as indicated by the sharp peak of the curve.

The super-imposed frequencies, producing the sidebands, cause

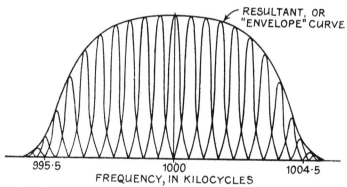

FIG. 83.—The resonance curve of a modulated broadcast transmitter. The speech currents superimposed on the carrier wave produce sidebands, or additional radiated frequencies, which cause the transmitted energy to be spread out over a wide band of frequencies.

the transmitted energy to be spread over a very wide band of frequencies, instead of, as in Fig. 82, being concentrated on a single frequency. The peaks of all the curves shown in Fig. 83 can be joined up, thus giving a resultant, or "envelope" curve which is taken to be the resonance curve of the transmitter, and receiver, during modulation.

If broadcasting stations on neighbouring wavelengths do not keep accurately to their assigned frequencies, or radiate sidebands more than 9 Kc. wide, the fringe or "skirts" of their resonance curves will overlap, as shown in Fig. 84, and audible interference will be set up in a wireless receiver tuned to the wavelength of either station.

Thus it is that television transmissions, using the broadcasting wavelengths, are at present limited in size (field of view) and detail to the amount of visual intelligence which can be conveyed through the medium of a 9 Kc. frequency band.

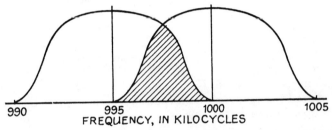

FIG. 84.—If two broadcast stations were separated by 5 Kc. instead of 9 Kc., the "skirts" of their resonance curves would overlap, thus creating interference in a receiver tuned to either station.

If we accept medium waves as being most suitable for television broadcasts, how, if at all, are we to overcome the difficulties in the way of broadcasting higher frequencies? The author has always been of the opinion that the answer lies in the direction of improved wireless transmitting and/or receiving apparatus. It is impossible to contemplate the contrast between Figs. 82 and 83 without coming to the conclusion that present technical methods and apparatus are extremely wasteful of what might be termed "wavelength space." That a telephony transmitter is inefficient might also be argued on the evidence that, taking a given transmitter, its unmodulated wave (Fig. 82 curve) has a range, if interrupted and used for telegraph purposes, of three or four times the distance over which it is possible to transmit intelligible speech when the carrier wave is modulated (Fig. 83 curve). The very nature of the two curves would lead one to expect this. When a broadcast listener, searching for distant stations with an oscillating receiver, comes across a strong carrier wave which, in spite of its apparent strength, he cannot " resolve " into audible speech or music, he has encountered an exemplification of the above argument.

Between 250 and 550 metres there is room for only 72 broadcasting stations to operate (on a 9 Kc. separation) without causing mutual interference, unless they are so widely separated geographically as to be out of range of one another. If we could somehow impress on a carrier wave all the intelligence we wanted to, either aural or visual, without causing a frequency spread of more than, say, 1000 cycles, we could build nine stations for every one in existence to-day, separating each station by 1000 cycles instead of 9000 cycles.

The situation is even more serious than this. Broadcasting is only one of the uses to which wireless is put for communication purposes, and the broadcast waveband is only a small part of the entire spectrum of wavelengths. Between about 15 and 26,000 metres thousands of wireless stations are packed in like sardines in a tin, all engaged on some important communication service or other, such as telegraphy, telephony, facsimile (phototelegraphy) transmission, military and naval services, broadcasting and experimental work. There is virtually no room for any more stations on any wavelength, for any purpose, under existing technical methods.

But have we reached finality in wireless technique? Is the sideband theory correct, or is it merely a convenient theory designed to explain observed phenomena in existing apparatus? The answer to the first question is that there is no such thing as finality in scientific matters; at the moment of greatest need the solution to an apparently insoluble problem is always forthcoming. The second question will be dealt with in what follows.

During the early part of 1930 there was considerable discussion on the subject of sidebands in the pages of *Nature*. This discussion was started by that veteran but still sturdy wireless pioneer, Sir Ambrose Fleming, F.R.S., who sought to show that sidebands do not, in fact, exist, but are merely a convenient mathematical fiction, designed to explain observed phenomena. In winding up the discussion, Sir Ambrose further suggested that sidebands do not exist in ether space, but are the creation of our only means of exploring that space (from a wireless point of view), i.e. our present wireless receivers. This discussion proved very highly controversial, one of the main arguments against Sir Ambrose's theory being the evidence of selective fading on short waves.

The question as to whether sidebands have a real existence or not was first brought to public attention in the autumn of 1929, when Dr. James Robinson announced his invention of a new type of wireless receiver, known as the Stenode Radiostat. At a public demonstration of the instrument in London later, at which the author was privileged to be present, it was shown to be so highly selective that the tuning of broadcasting stations was effected within a very few cycles. A local oscillator, tuned to within 100 cycles of the wavelength of either of the Brookmans Park stations caused absolutely no interference, and badly heterodyned (in an ordinary receiver) European stations could

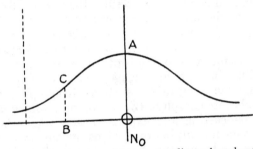

FIG. 85.—A typical resonance curve of an ordinary broadcast receiver.

be received entirely free of any interference. Moreover, this high degree of selectivity did not, as one would anticipate from experience of ordinary practice, impair in any way the quality of reproduction, which was as good as that obtainable from an ordinary good quality receiver.

Being very well aware of the overcrowded state of the ether, referred to above, Dr. Robinson commenced work some time ago to try and discover some method of squeezing stations closer together without creating interference, so that more channels could be made available for the host of new services which must inevitably seek space in the ether sooner or later, and for which there is at present no room. For the following particulars of the underlying principles and development of the Stenode Radiostat the author is indebted to Dr. Robinson.

In Fig. 85 is shown a typical resonance curve of an ordinary

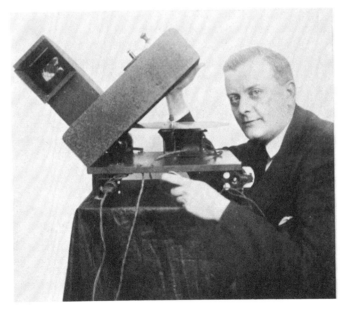

[*Photos Courtesy: Telehor Akt.-Ges.*

Above: The telecinema transmitter demonstrated by Mihaly at the 1929 Berlin Radio Exhibition. *Below:* Denys von Mihaly photographed with one of his 1929 model television receivers.

TRANSMISSION CHANNELS

wireless receiver. This curve is obtained by allowing continuous waves of different frequencies to arrive at a receiver, and by noting the current which is obtained for each particular frequency. The frequency which gives the maximum response (peak of the curve) is the frequency to which the receiver is tuned. The resonance curve therefore means that when a receiver is tuned to a certain frequency, it responds best to waves which possess that frequency; but it also responds to waves of a different frequency, the amount of response diminishing as the frequency of the incoming signals recedes from that to which the receiver is tuned.

Now consider a different case. The receiver is tuned to frequency N_0 and the incoming signals are of the same frequency, but are modulated. The sideband theory, by employing Fourier's theorem for the analysis of complicated waveforms, asserts that the modulated waves of frequency N_0 can be looked upon as a series of waves of different frequency, as shown in Fig. 83, but each of constant amplitude. However, one reality is that the waves which are arriving are of one frequency but of variable amplitude. Until recently (that is, until the development of the Stenode Radiostat and the discussion started by Sir Ambrose Fleming in *Nature*), it was becoming usual for wireless engineers to consider the reality to be that modulated waves actually do consist of a series of waves of different frequency, but each of constant amplitude. The new facts, however, have made engineers look upon the sideband theory in its correct light, which is that modulated waves may be regarded as a series of independent waves, although they are not necessarily so.

On the basis that modulated waves consist of a series of different waves which are all transmitted simultaneously, it is necessary to provide a wireless receiver having a resonance curve wide enough to receive the whole of these frequencies. Thus, for telephony, the resonance curve should be 9 Kc. in width, and, in fact, it should be more than this, because the resonance curve should be almost horizontal over the 9000 cycles in order that all frequencies may be received equally.

This implies that the resonance curve of normal type should actually be 20,000 cycles or more in width in order to receive one broadcast station without distortion. Thus, the sideband theory places a definite limit on selectivity and indicates that for telephony we must have a receiver with a resonance curve *at least* 9000 cycles in width, and for television a still wider resonance curve.

However, Dr. Robinson proceeded to examine what would happen if he employed a receiver with a resonance curve much narrower than that which was apparently demanded by the sideband theory.

In Fig. 86 are shown two resonance curves *a* and *b*, *a* being of the normal type for telephony, and *b* much more selective.

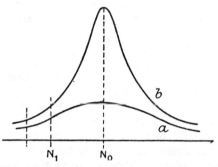

Fig. 86.—Two resonance curves, one normal (*a*) and the other (*b*) much more selective.

The frequency for correct tuning is shown as N_0 and a frequency of N_1 is shown which differs from N_0 by, say, 5000 cycles, so that curve *a* is suitable for telephony reception, as all frequencies which are supposed to take part in the carrying of telephony are more or less equally received. On the face of it, curve *b* would appear to give a large amount of distortion, and it was common to state that a circuit with a resonance curve of this type cuts off the upper sidebands.

Let us examine, however, what happens when signals arrive at any receiver. In this case, it is useful to observe how an incoming signal makes the receiver oscillate. The signals make the receiver oscillate at first with a low amplitude, which amplitude increases to a definite maximum. When the signal ceases, the receiver oscillations do not die down to rest instantly but continue for a time.

There are great differences in this response between circuits of high and low damping. In Fig. 87 are shown the build-up curves corresponding to the resonance curves *a* and *b* of Fig. 86. In Fig. 87 actual high-frequency oscillations are shown, but the feature which is of most importance is the actual envelope of these high-frequency oscillations, and these are shown in Fig. 88. From these curves we learn certain general principles; firstly, that the maximum amplitude which is attained increases as the resonance curve gets sharper; secondly, a definite period of time is required for the receiver to build up oscillations to

TRANSMISSION CHANNELS 147

their maximum value, and this build-up time is shorter the sharper the resonance curve; and thirdly, at the end of a signal, the dying-down period of the oscillations is greater in the case of the sharper resonance curve.

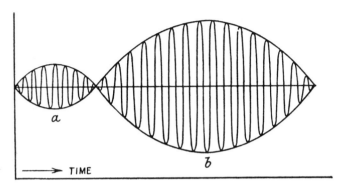

FIG. 87.—The build-up curves corresponding to the curves a and b in Fig. 86.

These three principles can easily be demonstrated by the reader himself, provided that he is a sufficiently experienced wireless amateur or experimenter. Take a wireless receiver fitted with reaction and tune in slow-speed telegraph signals. Now

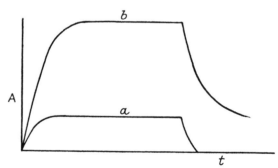

FIG. 88.—The actual envelope of the high-frequency oscillations shown in Fig. 87.

adjust the reaction until the threshold point of self-oscillation is almost reached, and it will be observed that each signal takes an appreciable time to build up to full strength, and a slightly longer time to die out. Better and more easily observable results will be obtained on wavelengths over 10,000 metres (the longer the wave the better), using an independent heterodyne (local oscillator) to make the C.W. signals audible.

The relation between this experiment and the argument at present being advanced is, of course, that the nearer the reaction is adjusted to self-oscillation point, the sharper the resonance curve of the receiver is being made. As all wireless enthusiasts know, a receiver in this condition is hopeless for receiving broadcasting; the resonance curve is so sharp that the most appalling distortion results.

It is very interesting to observe the effect of signalling speeds on circuits with different resonance curves, and in Fig. 89 the effect is shown for telegraphic dots, the dots being shown at P, Q, R, etc. It is obvious that for curve a, Fig. 86, each dot will cause the receiver response to build up to its maximum and die down to zero, and thus we get a separate response for each dot, as shown by curve a, Fig. 89. In curve b, however, at the end of the first dot the receiver has not had time to cease oscillating before the second dot arrived, so that we have a response curve, due to the dots, of a very complicated shape, and the receiver is tending to build up to an indeterminate value. This effect can also be demonstrated by the experiment quoted above. A point of reaction adjustment can be found, when no signals are arriving, when the arrival of two or three weak signals, or one strong one, such as a loud atmospheric, will set the receiver off into continuous self-oscillation.

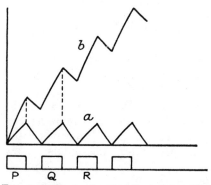

FIG. 89.—Illustrating the effect of signalling speeds on circuits with different resonance curves, such as a and b in Fig. 86.

Here we have a very important fact in electrical communication, which is to the effect that in order to obtain faithful reproduction, a suitable amount of damping must be introduced into a receiving circuit of the ordinary type. However, in view of the seriousness of the present congestion in the ether, it is essential to make an attempt to obtain this faithfulness of

TRANSMISSION CHANNELS 149

reception even when we have a very selective receiver, which receiver is made selective by removing the damping. It is of importance to note also that the deliberate introduction of damping makes a receiver less sensitive, by limiting the extent of the energy response, or build-up, which can take place under the influence of incoming signals.

In order to obtain selectivity (and increased selectivity) we remove the damping and once the receiver is excited by an incoming signal it takes a very long time to die down to rest or, in other words, its oscillations are very persistent, and they must be stopped, or quenched before the receiver is fit to respond to another signal. Advanced wireless enthusiasts who remember the old quenched spark discharger will understand what is required; one function of the quenched spark discharger was to quench oscillations in the primary circuit of a transmitter, so as to leave the aerial circuit free to oscillate at its own frequency without interruption from subsequent oscillations or impulses from the primary circuit.

It is therefore essential to try to find some means for overcoming this persistence of very selective receivers, some quenching device; and when we have done so, we obtain the result that we can get faithfulness of reproduction of signals (telegraphic, telephonic, etc.) in conjunction with the fact that the receiver still has a very sharp resonance curve, so that it can only be excited appreciably by signals which lie within a comparatively narrow band of frequencies. The Stenode Radiostat has provided the solution to this problem.

One method by which this has been achieved is to cause signals to act on a very selective receiver, first of all to build up the receiver oscillations to a certain value, which is not necessarily their maximum possible value, and then to cause the same signal to bring the receiver to rest again. It is necessary to make this occur at a frequency which is higher than any frequency which takes part in the actual modulation.

Thus, if we are concerned with telephony, the receiver must be excited and brought to rest more than 5000 times per second. This effect is achieved by causing incoming signals to act on a receiver, first of all in one phase and then for a similar period in the reverse phase.

An elemental circuit for achieving this phase reversal is shown in Fig. 90, where incoming signals are applied in opposite high-frequency phase to the grids of two valves. There is one other grid in each of the valves, and when these two grids are of the same potential, no high-frequency signals can pass through the combination, as we have a state of balance. However, a low-frequency oscillation is applied to the extra grids in such a way that these two grids are always at opposite potentials.

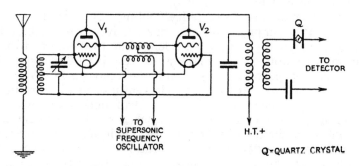

FIG. 90.—An elemental circuit for quenching by the phase reversal method excessive build-up in a selective receiver without increasing the damping.

In the case of telephony the frequency of the oscillation would be of the order of 10,000 cycles per second. From the parallel-connected anodes of this combination, signals are fed to the selective receiving circuit through any convenient form of coupling.

The result obtained from such a combination is shown in Fig. 91, from which it is seen that we obtain pulses of high frequency, the number of pulses being the same as the low frequency which is fed to the second grids of the valves in Fig. 90. The amplitude of these separate pulses in Fig. 91 obviously depends on the instantaneous amplitude value of the incoming signals, so that if these signals are varying in amplitude, we find that we obtain an envelope of response which, after rectification, corresponds to the original signals which were transmitted.

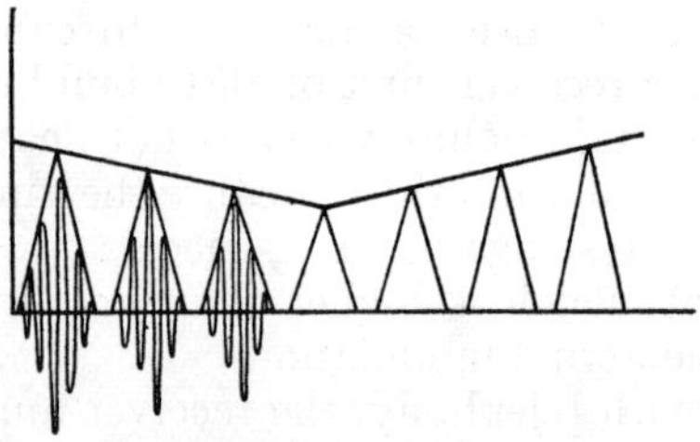

FIG. 91.—Envelopes of high-frequency currents obtained from the Fig. 90 combination.

[*Courtesy: Telehor Akt.-Ges.*

The Telehor Company's (Mihaly) 1930 model universal television receiver, designed to receive either the German or British television broadcasts. Views show both external and internal arrangements.

[*To face page* 150.

It is obvious that such a receiver can only be excited by signals which are within the narrow resonance curve which is employed, so that if we use a receiver with a resonance curve less than 100 cycles in width, we have a highly selective receiver, and we can still obtain perfect reproduction of modulated signals. Actually, the selective arrangement which is employed is a piezo-electric, or quartz crystal device, and the resonance curve of the entire combination is well below 100 cycles in width.

This means, therefore, that since the Stenode Radiostat can efficiently receive a B.B.C. station on a frequency band only 100 cycles wide, the remaining 8900 cycles allotted to the station represent just so much waste space in the broadcast waveband. In other words, on the 9 Kc. frequency band where *one* broadcast station now operates there is room for a total of *ninety* stations, and by using one of the new receivers having a resonance curve just under 100 cycles wide, one could pick them all up perfectly, without interference.

It is understood that Dr. Robinson proposes to apply his new principle to transmission as well, with the object of designing a transmitter which, when received by means of a Stenode Radiostat, will, amongst other advantages, enable the broadcasting of much higher modulation frequencies than it is possible to transmit at present, and without creating further interference. There are excellent grounds for believing, therefore, that the problem of communication channels for television is not altogether insoluble.

Another difficulty in connection with the broadcasting of television is that two wireless transmitters, working on two different wavelengths, are necessary, one to transmit the image signals and one to transmit the accompanying speech or music produced by the artiste sitting before the television transmitter and microphone. A way of overcoming this difficulty, known variously as multiplex radio, double modulation, or the Hammond system, consists of superimposing both the image signals and the speech signals on the same carrier wave. This system was experimented with by the Bell Telephone Laboratories and the methods used, and the results obtained, were described in the October, 1927, issue of the *Bell System Technical Journal*, upon which the following description is based.

The output of a 30,000 cycle oscillator was modulated with the speech signal. The resulting carrier and sidebands were selected by means of a suitable filter passing frequency components ranging between 25,000 and 35,000 cycles (a 10 Kc. band) and impressed on the input terminals of a wireless transmitter along with the 10 to 20,000 cycle image signal from the television transmitter. A suitable low-pass filter was employed in the line to the latter in order to preclude "cross-talk," or interference due to 25,000 to 35,000 cycle energy from the speech circuit working back into the final amplifier stages. The input to the wireless transmitter thus consisted of two frequency bands, one extending from 10 to 20,000 cycles, together with a 25,000 to 35,000 cycle band, with a particularly strong component at 30,000 cycles representing the low-frequency speech-current carrier.

In order that it might be capable of handling this extremely wide frequency band without discrimination, the wireless transmitter employed had to be considerably modified. In the case of some of the radio-frequency circuits, which were required to pass a 70,000 cycle band, it was found necessary to increase the damping by inserting resistance, thus reducing the sharpness of the resonance curve.

It is well known that a modulated current, or wave, may be resolved into three components : (1) a steady current, known as the carrier, which has the amplitude and frequency of the original unmodulated current, (2) an upper sideband (right-hand half of Fig. 83) which is equivalent to the signal spectrum, or range of input frequencies, with each individual frequency increased by an amount equal to the carrier frequency, and (3) a lower sideband (left-hand half of Fig. 83) which is an inverted reproduction of the signal spectrum, that is, each individual signal component is laid off in the downward direction from the carrier frequency, or subtracted from it. Hence, assuming a carrier frequency of 1575 Kc. and a signal input to the wireless transmitter of the type described above, the aerial current, or the transmitted wave, may be represented diagrammatically as shown in Fig. 92.

It is evident that this type of wireless signal can be received by employing an arrangement which will accept the entire 70 Kc. band, and subject it to rectification in the usual manner. If this is done, the television signal, together with the 30 Kc. supplementary carrier modulated with speech, will appear at the output of the detector. Branch circuits with suitable filters will then enable these two components to be separated,

TRANSMISSION CHANNELS

and the television signal passed on to the reproducing apparatus. The other component must be rectified again to derive the original speech signal, which may then be impressed on the loud speaker amplifiers.

The reception scheme actually adopted during the Bell experiments was somewhat different. The television signal was received separately by means of a special triple detection receiver (a specially highly selective form of superheterodyne)

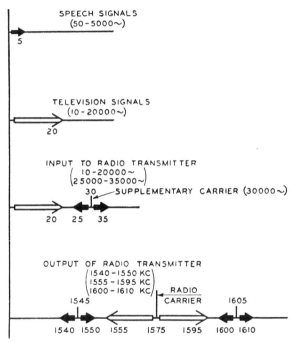

FIG. 92.—Diagrammatic representation of frequency conversions in a multiplex wireless system.

tuned to 1575 Kc. The speech signal was received on a receiver of conventional design, tuned to 1545 Kc. That reception in this manner is feasible is evident from Fig. 92. The 1540-1550 Kc. zone contains two speech sidebands and a carrier of $1575 - 30 = 1545$ Kc. It is quite possible, therefore, to demodulate, or rectify in one step, instead of " beating " the various components against the main carrier (1575 Kc.) to produce a 30 Kc. supplementary carrier which must be rectified a second time to derive the speech signal. The 1600-1610 Kc. band was ignored. The receivers were sufficiently selective that, with the 5 Kc. interval which separated the two bands, speech and television, no noteworthy interference was experienced.

The results obtained in this manner were not as satisfactory as those to be had with the system first described, where the entire 70 Kc. band is received on one receiver. This was attributed to two factors, both concerned with the transmitting apparatus. In the first place, in order to transmit both signals with the same transmitter, i.e. the same transmitting valves, the individual current amplitudes had to be reduced to at least half in order to prevent the combined signals causing overloading or " blasting " of the valves. This reduction in individual amplitudes resulted in too weak a received signal to clear the prevailing interference noise level in New York, where the receiving equipment was situated. In the second place, in spite of the reduced amplitudes, a certain amount of intermodulation was experienced in the transmitter which resulted in interference between the channels. Notwithstanding these deficiencies, however, it was possible to recognise the speaker and to understand his remarks.

Experiments of this nature, although not new, are of particular interest where television is concerned, since the logical trend of development is towards a finer picture structure involving the transmission of much wider frequency bands, or—what the Bell Telephone engineers regard as more likely—the use of parallel scanning schemes and multi-channel transmission. From a popular standpoint, the experiments have been described as the transmission of both voice and image " on a single wavelength." To what extent this statement falls short of actually representing the facts of the case is obvious from Fig. 92. It will be seen that a wider frequency band (70 Kc.) is actually employed with this system than is required for two separate channels, which would occupy 10 Kc. for speech and 20 Kc. for television, or a total of 30 Kc.

Furthermore, this wider band is much less effectively utilised. Two bands are required for the voice channel in place of one, and at the receiver one of these bands was disregarded. To have received both would have required apparatus accepting twice the band width, and the gain in signal strength would have been offset by the corresponding increase in noise level. For all useful purposes, therefore, the energy radiated in the form of the second band was wasted.

An economy in band width could be effected of course, but at the expense of greatly increased complication at both transmitter and receiver, by making use of the single-sideband, carrier-eliminated transmission system originally developed by the Bell Telephone Laboratories and now in use on the long-wave transatlantic wireless telephone circuit. This system, as the name indicates, does not transmit the carrier wave, and further economises in energy and frequency bands by radiating one sideband only. As the second sideband is but a duplicate of the other, it can be re-introduced at the receiver by modulating a locally generated carrier wave with the single incoming sideband. It will be interesting to see what combination is arrived at between these existing schemes of transmission and reception, and the Robinson system.

The Bell Telephone engineers, in concluding their paper on multiplex transmission, state that "the frequently predicted introduction of television as an adjunct to radio broadcasting without extensive changes in existing channel arrangements is extremely unlikely." This expression of opinion, coming from such a source, cannot be ignored, and might well have provided the text for the discussions contained in this chapter.

In addition to their work in connection with wireless channels for television, the Bell engineers have also carried out extensive investigations in the transmission of television signals by wire circuits, especially in cable form. Their conclusions as to the relative values of the two mediums of communication read as follows:

" The requirement of an extremely wide transmission band, and the further requirement that during the period of transmission the channel or channels must have a high degree of electrical stability and freedom from extraneous interference, make the channel problem both difficult and expensive. Unlike telegraph or telephone transmission, where a limited amount of channel instability or a moderate amount of electrical interference can be present without serious impairment of service, phototelegraphy and particularly television require practically perfect interference—free channels. For these services any marked instability in the channels or any substantial electrical interference registers at once as a serious defect in the received image.

" It is for this reason that while radio channels, if otherwise available, can be used for the transmission of television, they are not in the present state of the art as suitable as wire channels. Wire telephone circuits, particularly if in cable, can be maintained at a high degree of constant transmission efficiency and freedom from extraneous interference. Radio channels, on the other hand, are subject to the well-known vicissitudes of fading, static and interference, all of which result in a degraded received image."

[*Courtesy: Baird Television Ltd.*

The Baird Company's first television broadcasting studio at 133 Long Acre, London, from which the first regular experimental television broadcasts were transmitted. Note microphone, to the right of which is the opening in wall through which the light spot comes from the control room. Above the opening can be seen the photoelectric cells.

157

CHAPTER XII.

THE PRESENT STATE OF THE ART IN GERMANY.

Fernseh, A.G., Arrangements. German Post Office Apparatus. Aims of Post Office. Karolus-Telefunken System. Telefunken Short-Wave Experiments. The Mihaly System. Von Ardenne's Experiments. Review of Position in Germany.

IN Germany the principal television workers are Denys von Mihaly, a Hungarian who has already been referred to earlier in this book ; Prof. Karolus of Leipzig University, who works in conjunction with the Telefunken Company ; the German Post Office ; Fernseh, A.G., an offshoot of the British Baird Company ; and Manfred von Ardenne, a well-known radio engineer.

The German Post Office first began to display an active interest in television in 1928, when it commenced to co-operate with Mihaly by subsidising and otherwise assisting his experiments. Early in 1929, after several of the German Post Office engineers had witnessed a demonstration of the Baird system in London, official interest in Baird became so keen that the Post Office assisted in the preliminaries of the flotation of Fernseh, A.G., which is an equal partnership between the Baird Company, Zeiss-Ikon (optical products), Bosch (magneto and electrical equipment), and Loewe Radio.

The Post Office also commenced active experiments of its own, based on the Baird system, but designed to investigate the problems and possibilities of the broadcasting, not of television, but of cinematograph films. For this purpose they chose a 30-hole disc, horizontal scanning, and a picture ratio of 3 to 4. At the transmitter, films were run through a standard cinema projector, the light beam being interrupted by a scanning disc which scanned each " frame " of the film.

The receiver followed the design of the Baird system, but instead of using the usual flat plate neon tube as a light source, a rather large mercury-argon light source was employed. This consisted of a thin glass tube bent to and fro upon itself to form a grid measuring about 4 inches by 6 inches. Between this grid and the back of the scanning disc was placed a ground glass screen to diffuse the bars of light from the grid-shaped tube. In front of the disc, nearest the observer, was placed the usual large magnifying lens.

This receiver, which was demonstrated at the 1929 Berlin Radio Exhibition, made no pretensions to being commercial, for it was too large for home use, and the mercury-argon lamp required something in the neighbourhood of 200 watts (obtained from a large power amplifier) to operate it. The light from such a tube is bluish in colour, and the picture, though good in detail, was lacking in depth and intensity of illumination.

The speed of transmission standardised by the German Post Office is the same as that used by Baird, $12\frac{1}{2}$ pictures per second. This speed is too low, for it produces a most unpleasant flicker effect, but it has the advantage that it reduces the transmission frequency, an important factor when it is remembered that for the present we have to transmit as much visual information as possible on a 9 Kc. frequency band.

Another feature of the Post Office exhibit at the 1929 Radio Exhibition was a two-way television system. Two telephone boxes were erected at opposite ends of the Post Office enclosure, in which a member of the public could sit and converse, by telephone, with another person seated in the other box, and the two could simultaneously see one another.

For this purpose a single scanning disc was used at each end of the circuit, so arranged that while the face of the telephonist was being scanned by a beam of light emanating from the holes at the bottom of the disc, by looking at the top of the same disc he saw there the image of the person at the other end of the wire. The received image, however, was not at all good and, owing to the spot-light playing on one's face, it was difficult to see.

The aim of the German Post Office is to perfect the best possible form of television transmitter, incorporating the ideas and patents of all-comers. To this end experiments in the broadcasting of the films have been made daily from Witzleben, a $1\frac{1}{2}$ Kw. broadcasting station on the outskirts of Berlin, since 1928. More recently, a suitable land line having been installed, experimental television broadcasts are being made from the powerful long-wave station at Koenigswusterhausen, a few

PRESENT STATE IN GERMANY

miles outside Berlin. The German Post Office controls all sound broadcasting in Germany, and its object in taking such a practical interest in television is to assist in perfecting it to a point where it can be broadcast regularly, under Post Office control, in conjunction with sound broadcasting. The Post Office is not concerned with the development of receivers; that is being left to the enterprise of commercial firms who will market them as soon as television broadcasts come to be transmitted regularly on something more than an experimental basis.

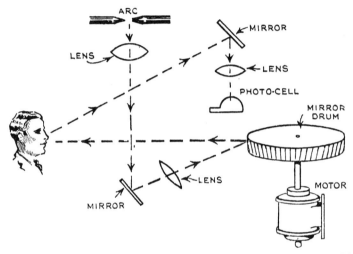

FIG. 93.—Diagrammatic representation of the Karolus-Telefunken television transmitter.

The Karolus-Telefunken system has many original and interesting features. In the first place a Weiller mirror wheel, or drum, is used instead of the usual Nipkow scanning disc. The layout of the transmitter is shown diagrammatically in Fig. 93, and photographs of the transmitter itself are reproduced, facing pages 120 and 126.

A beam of light from an arc lamp is concentrated by means of a lens system to a fine and extremely intense point, which is directed on to the revolving mirror drum. As has already been explained in Chapter II., when light falls upon a mirror the angle of reflection is equal to the angle of incidence. In the system shown in Fig. 93, the incident angle of the beam of light from the arc is always the same, but it is desired to constantly vary the angle of the beam reflected from the mirrors which are set on the periphery of the drum, in order to move the beam up and down and thus achieve the complete scanning of the face of the sitter. This is achieved by slightly staggering each mirror relatively to its neighbour. The mirrors then impart vertical movement to the beam, while the revolving motion of the wheel imparts horizontal movement.

From the mirror drum the light beam is reflected on to the face of the sitter, when it is again reflected and concentrated by means of a lens on to a single photoelectric cell. The use of one cell only places the Karolus-Telefunken system at a disadvantage by comparison with most other systems which use a bank of several cells connected in parallel, for, in order to obtain an adequate photoelectric response, Karolus finds it necessary to use a much more powerful scanning beam, to the consequent discomfort of the sitter. The author found the beam to be most unpleasantly dazzling. A considerable increase in the sensitivity of the photoelectric cell and/or the associated amplifiers is the only way in which this difficulty can be overcome without radically altering the layout of the transmitter.

It should perhaps be explained that the tortuous course taken by the light beam in Fig. 93 is not essential ; by so arranging matters it is possible to encompass the transmitter in a smaller space than would otherwise be possible if the focal lengths of the various optical systems were arranged in straight lines.

The number of mirrors mounted on the mirror drum is forty-eight, and the picture is square, so that the number of picture elements is 2300. The transmission speed at present in use is twenty pictures per second, so that the transmission frequency amounts to 23 Kc.

To secure synchronism Dr. Karolus uses at the transmitter an electrically driven tuning-fork that provides a steady alternating current, which, after amplification, is fed to a small synchronous motor, which, in turn, is mechanically coupled to the main motor which drives the mirror wheel. This synchronous motor keeps the speed of the main motor constant.

At the receiving end an exactly similar tuning-fork provides alternating current for a synchronous motor which controls

PRESENT STATE IN GERMANY

the speed of the motor used to drive the receiver mechanism. As has already been explained, the weakness of this system lies in the tuning-forks which, together with the associated equipment, are not only expensive but not accurate enough for the high-speed requirements of television. Where a separate channel of communication is available, however, impulses from the transmitter can be sent direct to the receiver, thus eliminating the fork apparatus at that end of the circuit, and the control of both transmitter and receiver is then under one fork, that at the transmitter, and excellent results can then be obtained. Unfortunately, however, one transmitter may be required to serve a thousand receivers, and the provision of a thousand wire synchronising channels becomes an impossible problem economically.

Up to the present Karolus has produced two distinct types of receiver, a reasonably small one for home use and a larger one for use in small halls. The small receiver produces an image about 4 inches square, while the large instrument projects on to a screen an image about 3 feet square.

In the small receiver a thin pencil of light from the light source, a crater-type neon lamp, is focused on to a small mirror wheel measuring about 6 inches in diameter, whence it is reflected on to a ground glass screen. The image produced by this receiver is very good indeed as far as detail is concerned, but the choice of ground glass for a screen is scarcely a happy one. Not only does ground glass cause loss of light (illumination of screen), but it also causes the illumination of the screen to be very directional, i.e. if the observer stands in such a way that his eyes are dead level with the centre of the screen he can see the middle of the image quite clearly, but scarcely anything of the edges. In order to see the edges he must move his head bodily from side to side and up and down.

The large receiver operates on an exactly similar principle, but the light source, and the method of controlling it, are entirely different. A sketch of the arrangement is shown in Fig. 94. Light from an arc lamp is concentrated in an intense beam which passes through two sets of nicol prisms between which is situated a Kerr cell.

As is well known, the action of a nicol prism is to polarise a beam of light passing through it, i.e. cause the light to vibrate in one plane instead of in every conceivable plane. The first prism is arranged to polarise the light beam on one plane, while the second prism is arranged to polarise it in a plane at right angles to the plane of first polarisation. The net result is that the light vibrations in the two opposing planes cancel each other, and no light at all emerges from the second prism.

The Kerr cell (sometimes called the Karolus cell) consists essentially of two tiny condenser plates immersed in nitro-benzol. When a polarised beam of light passes through the liquid, and between the condenser plates, a difference of potential (electrostatic force) applied to the condenser plates tends to depolarise or twist the plane of polarisation of the beam of light. This is known to physicists as the Kerr effect, so named after its discoverer.

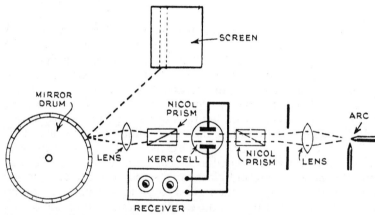

FIG. 94.—Diagrammatic sketch of the arrangement of the Karolus-Telefunken large screen receiver.

Referring to the Fig. 94 arrangement, it will readily be appreciated that, since a state of balance has been arrived at where no light passes the nicol-prism-Kerr-cell combination, any difference in voltage applied to the plates of the Kerr cell will partially depolarise the light beam and permit some light to pass through, and the amount of light which gets through is proportional to the applied electrostatic force. Since the action is instantaneous the arrangement would appear to be an ideal light valve, or control device, for controlling the intensity of an arc lamp of any desired degree or power, for all we have to do is to apply the incoming television signals, after amplification, to the plates of the Kerr cell. Unfortunately, however, the efficiency of such an optical system is very

[*Courtesy: Baird Television Ltd.*

Behind the scenes in the control room of the Baird Company's studio, showing the projector lamp, scanning disc, and "A" amplifier of the television transmitter.

[*To face page* 162.

PRESENT STATE IN GERMANY 163

poor, and an immense amount of light is lost. In the Karolus apparatus the nicol prisms are water-cooled, owing to their close proximity to the intense heat of the arc.

After passing through the optical system the beam of light, varying in intensity in exact accordance with the fluctuations of the incoming television impulses, is focused on to a mirror drum, whence it is reflected on to a specially prepared screen of a type exactly similar to that used in cinematograph theatres. Such screens are covered with finely powdered aluminium to increase the apparent brilliance of the picture.

The detail of this large picture is very good indeed, but extremely difficult to see, owing to the poor illumination which results from the use of the nicol-prism-Kerr-cell system.

During the past year or so, no important changes have been reported in the Karolus apparatus. Instead, the Telefunken Company has concentrated on problems in connection with the wireless transmission of television on short waves, using a 48-line transmitter sending twenty pictures per second.

The Telefunken Co. attacked the problem on the following premises. Reviewing the European broadcasting situation in general, it is argued that an ideal arrangement would be to eliminate all the present low-power broadcasting stations, and substitute for them one or two central transmitters in each country having great power and wide frequency separation. If such a scheme could be carried out, then it would be possible, technically and otherwise, to broadcast television signals having frequencies of 23 Kc. or even higher, so that they could cover the whole nation.

Since this scheme is not at present possible, the alternative appears to be to establish in the large cities local short range transmissions on short waves, increasing the detail of the picture, and thus the frequency, even more. On account of the tendency of short wave radiations to concentrate most of their power into free space waves, and very little into ground waves, serviceable locally, the range of such transmissions would be limited to the horizon of the transmitter. Even then, there is always present a tendency for interference to take place between space and ground waves, which causes fading and distortion to sound signals, and the production of multiple images in a television receiver.

The Telefunken Co. have been experimenting between Nauen and Geltow, which are 40 Km. apart, on a wavelength of 70 metres, and by a suitable design of both transmitting and receiving aerials have succeeded in screening off the space wave, so that only the ground wave is received at the receiver. In this way excellent images have been received, entirely free from multiple image producing "echoes," and it is stated that the apparatus employed is suitable for use with images having a degree of detail up to 96 lines. Images transmitted experimentally from Schenectady to Leipzig have also been received successfully, and even recorded on cinematograph film with good results.

Following the successes achieved by Baird in England by the use of the Nipkow disc system, Denys von Mihaly abandoned his oscillograph apparatus, described in an earlier chapter, and also turned his attention to the disc. By the autumn of 1929 he had produced two types of receiver for home use, which he proposed to market immediately, both in Germany and in England, and a larger machine for demonstration purposes. The author had the privilege of seeing these machines in operation at the 1929 Berlin Radio Exhibition.

Of the two machines designed for home use, one was very small and cheaply constructed. It consisted of a light ebonite disc driven by a small phonic wheel, the stator of which was supplied with current from the same A.C. mains which supplied the transmitter. The number of holes in the disc was only twelve. The light source was a small neon tube reported to consume only 10 Ma. At the demonstration witnessed by the author, this instrument was receiving the image of a lantern slide only, and the picture was of very poor quality, due to the small number of image strips. The instrument as a whole could only be regarded as a toy.

The larger receiver was somewhat better constructed. The disc, in this case a 30-hole one, was also driven by a phonic wheel, used as an induction type synchronous motor, the rotor of which contained thirty iron strips. The stator was supplied from the output terminals of a 3-stage amplifier, the input terminals of which were supplied with a 375 cycle synchronising current sent by separate land wire from the transmitter. The circuit connections to the phonic wheel are given in Fig. 95.

The voltage on the last stage of the amplifier was 150-200, and the output current 30-40 Ma. In series with the stator winding was a variable 1200 ohm resistance for rough speed adjustment. According to information supplied to the author,

PRESENT STATE IN GERMANY

the wheel will run on as low as 8 watts, but not very well. The resistance was generally set to supply the wheel with about 12-14 watts.

The neon tube employed consumed 50-60 Ma., and was supplied by another amplifier the input of which was the incoming picture signal. Only cinematograph films were demonstrated on this machine, but the reproduction was quite good.

In the cases of both these receivers, the motor shafts were mounted vertically, so that the position of the scanning discs was horizontal, with the neon lamp placed underneath. Viewing of the image was accomplished by means of a kind of periscope arrangement, the received image being reflected by a mirror set at an angle of 45 degrees at the top of a short vertical box arrangement, at the top of which, in front of the mirror, was set a magnifying lens.

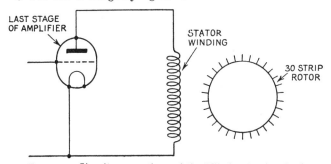

FIG. 95.—Circuit connections of the Mihaly phonic wheel.

The large demonstration receiver was mounted in a large wooden case measuring about 6 feet high. Vertical scanning and a large neon tube were employed. The disc, measuring about 3 feet in diameter and containing thirty holes, was driven off the A.C. mains by a synchronous motor which appeared to be of about $\frac{1}{4}$-h.p. rating. No separate synchronising mechanism was incorporated in this machine, the A.C. mains, which also supplied the transmitter motor, being relied upon to keep the receiving motor in step.

This machine also was used for the reception of cinema films. The image produced, as seen through a magnifying lens, appeared to be about 4 inches square. The brilliancy of the image was good, as was also the detail.

Mihaly did not demonstrate actual television at the 1929 Berlin Radio Exhibition, nor did he demonstrate any form of automatic synchronism, independent of A.C. mains, but he exhibited a diagram explaining how he proposed to accomplish it. This is illustrated in Fig. 96.

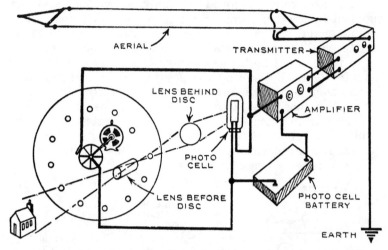

FIG. 96.—Diagram of an automatic synchronising system proposed by Mihaly.

A commutator mounted on the shaft of the transmitter driving motor is arranged momentarily to short-circuit the

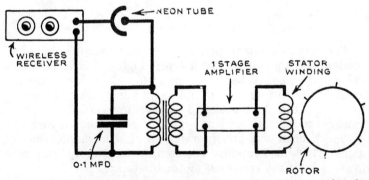

FIG. 97.—Receiving circuit arrangements for use with the proposed system of synchronising shown in Fig. 96.

photoelectric cell six times per revolution, so that for every revolution of the motor six very strong impulses from the photo cell battery would be sent to the receiver. It may be

PRESENT STATE IN GERMANY

remarked in passing that such treatment is scarcely designed to improve photoelectric cells and their associated batteries.

The receiver arrangements for use with this system are shown in Fig. 97, where the output of the wireless receiver, which comprises both picture and synchronising impulses, is fed first to the neon lamp and then through the low impedance primary of a special transformer. Across this primary is con-

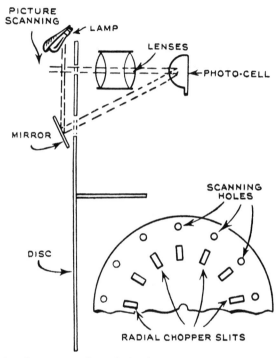

FIG. 98.—Another proposed method of generating synchronising impulses at the transmitter.

nected a ·1 mfd. condenser to by-pass the high-frequency picture component of the amplifier output.

The output of the transformer secondary is then taken to a single stage amplifier, the output of which should then be sufficiently powerful to run the phonic wheel which drives the receiver scanning disc. The rotor of the phonic wheel, in this case, would have six strips.

Such a method of synchronism would produce a very bright strip or band either at the top or bottom of the received image.

Another proposed method of generating synchronising impulses for transmission to the receiver, together with the picture impulses, is illustrated in Fig. 98. This system was designed for use with a cinematograph film transmitter, not a television transmitter. In this system the scanning disc, in addition to being pierced with the usual scanning holes, is equipped with an equal number of radial slots, set nearer the centre of the disc, as shown. Mounted above and behind the disc is a powerful lamp, the rays of which are deflected through the radial slots and on to the photoelectric cell by means of a mirror. It was claimed for this arrangement that, at the receiver, a white strip would be produced at the top of the image which would result in the loss of no more than 5 per cent. of the picture. However, it appears to the author that the powerful flashes falling on the cell from the lamp would tend to mask out the scanning beam altogether and produce at the receiver nothing but a brilliantly lit blank screen.

Manfred von Ardenne has come into prominence quite recently in connection with television research. He has broken away from mechanical methods entirely, and is experimenting with cathode ray tubes of the incandescent cathode type. One of the difficulties in connection with cathode ray schemes is to obtain a sufficiently high degree of brilliancy from the fluorescent screen, upon which the cathode ray stream impinges and becomes visible.

Von Ardenne claims that the best results have been obtained with green fluorescence, because green is the colour to which the retina of the human eye is most sensitive. It is desirable also that the luminescence of the screen should persist, and after much experimental work he selected a fluorescent material containing a certain amount of silicate of zinc, which gave good results.

In the ideal television screen the brilliance of the image should at least be equal to that of the average cinema screen. Measurements of the von Ardenne fluorescent screen show that the degree of brilliance obtained even allows of magnification of the size of the pictures, for the light intensity, as measured, amounted to 240 Hefner candles per square metre, which is more than twice the usual intensity of illumination on the projection screens of large cinematograph theatres. It is, therefore, possible to observe the images in full daylight.

No further details concerning von Ardenne's work are

[*Photos Courtesy: Baird Television Ltd.*

External and internal views of the Baird "Televisor" receiver.

[*To face page* 168.

PRESENT STATE IN GERMANY

available at the time of writing, but it would appear that his experiments up to the present are in the nature of a preliminary survey of the field and its possibilities.

At the time of writing there are no commercial televisors on the German market. At the 1929 Radio Exhibition, Mihaly announced that he would have instruments similar to the second of his types, described above, on the market by the beginning of 1930, at a price between £10 and £15. His plans did not materialise, however. At the same time Fernseh, A.G., announced that the first 300 of their Baird "televisors" were in production, and that they would be on sale before the end of 1929 at a price equivalent to about £17 10s. These plans did not materialise either. The Telefunken Co. stated definitely that they had no intention of marketing television receivers at that time, and the same policy still holds.

At the 1930 Berlin Radio Exhibition, Fernseh exhibited receivers very similar to the 1929 model. Mihaly exhibited a "universal" disc-type receiver, suitable for the reception of either the British or German transmissions, and incorporating a phonic wheel synchroniser similar to that now used by Baird (see next chapter).

As things stand at the time these lines are being written, it appears to the author that Germany must look to the Telefunken Company, working in conjunction with Dr. Karolus, to develop television into a really practicable proposition, so that instruments can be sold to the public which will provide acceptable entertainment. And the Telefunken Co. definitely recognises the limitations of all mechanical methods of scanning, and seeks to develop the cathode ray tube as a scanning and projection means, both at the transmitter and receiver.

CHAPTER XIII.

THE PRESENT STATE OF THE ART IN ENGLAND.

Television Broadcasting Facilities. Baird Studio and Control Room. Baird "Televisor" Receiver. Cog Wheel Automatic Synchroniser. Relay Synchroniser. Baird's Large Screen. Tele-Talkies. Correcting Faulty Reception. Echo Images. Review of Position in England.

Up to the time of writing, there has continued to be but one prominent television experimenter in England, J. L. Baird, whose company, Baird Television, Limited, is at present engaged in an attempt to commercialise television in its present form. Early in 1930 the company placed upon the market a receiver the trade name of which is the Baird "Televisor" Receiver, at a cost of £26 5s. A kit of parts was also placed upon the market, costing £16 16s.

The television broadcasting which these receivers can pick up is limited to a half-hour period from 11 to 11.30 a.m., Monday to Friday inclusive, and from midnight to 12.30 a.m. on Tuesday and Friday nights. These transmissions have their origin in the studio of the Baird Company at 133 Long Acre, London, W.C. 2. From there the outgoing signals, both sight and sound (for the transmissions are dual), are sent by landline to the B.B.C. twin stations at Brookmans Park, the television signals being transmitted from the 356 metre Regional transmitter, while the accompanying speech is sent out through the 261 metre National transmitter. The times of these transmissions are, of course, highly inconvenient for most people, and too brief to enable the amateur to experiment very much. Whether the times will be altered, or the periods of transmission extended, remains to be seen; for the present the B.B.C. emphasises that the transmissions are purely experimental, and will not be extended or put on a regular service basis until there is a sufficiently insistent public demand for such an extension of facilities.

PRESENT STATE IN ENGLAND

Apart from the Baird B.B.C. transmissions, the only other source of television signals is Germany, and since the Germans, as already explained, use horizontal scanning and Baird uses vertical scanning, the reception of the German transmissions is a matter of some difficulty. Once the image has been picked up it is necessary to incline one's head sideways at an angle of 90° in order to recognise, or attempt to recognise, the picture. Thus, conflicting standards are bringing confusion to a new art at the outset. If the Germans also used a disc with a different number of holes, it would be quite impossible to pick up their transmissions on a Baird receiver.

A photograph of the interior of the Baird Company's first studio is shown facing page 156. It is similar to the usual form of draped broadcasting studio, with the exception that there is a small rectangular opening cut through one wall. The artist sits in a chair in front of this opening, through which the scanning beam passes from the control room and plays on his face. Light reflected back from the sitter's face strikes the photoelectric cells, which are mounted in a shielded case at the top of the opening. Four cells are used in parallel, and they are of the latest cæsium type developed by the G.E.C. The wall opening is fitted with a pane of glass, to prevent sounds from the control room affecting the microphone, and sounds from the studio affecting amplifiers in the control room and producing microphonic noises. Even with these precautions, if the artist speaks or sings too loudly, the photoelectric cells themselves vibrate mechanically, and these vibrations set up microphonic effects which can be seen in the receiver in the form of a shower of whirling white spots which, if very bad, will totally obscure the image. Fitted on to the photoelectric cell box there is a small mirror, so that the artist can see himself and determine whether he is properly seated within the scanning beam, neither too high nor too low, and exactly central.

The general layout of the control room, as originally arranged, is illustrated in Fig. 99. The output from the photoelectric cells goes to what is termed the " A " amplifier, thence to suitable switches which connect the amplified vision signal to either of two amplifiers called AX1 and AX2, the first of which is used for line demonstrations within the building, and the second for broadcasting. The circuit diagrams of these amplifiers are given in Fig. 100. The output of AX2 goes to a line corrector which is

172   FIRST PRINCIPLES OF TELEVISION

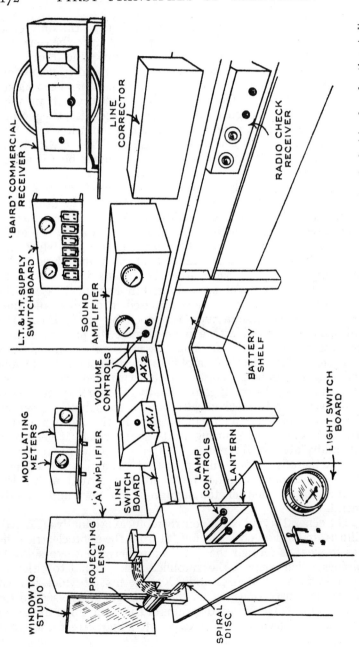

Fig. 99.—The arrangement of the control room adjacent to the Baird Company's sound and television broadcasting studio.

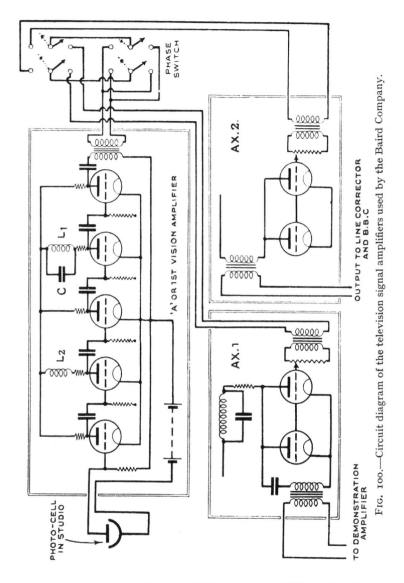

Fig. 100.—Circuit diagram of the television signal amplifiers used by the Baird Company.

intended to "balance" the land-line to Savoy Hill, so that it will transmit the vision signal undistorted. From the line corrector the signals go direct to Savoy Hill, where they are amplified, and sent via another line corrector to Brookmans Park. The output of the microphone, carrying the sound signal, passes through an amplifier which has two outputs, one to loud speakers within the building, and the other to a second line corrector, and thence to Savoy Hill over a separate land-line.

The leads connecting the photoelectric cells to the "A" amplifier must be made as short as possible, partly because the self-capacity of long leads has a serious shunting effect at the higher frequencies, and partly because long leads tend to pick up interference which would subsequently be greatly amplified along with the signal. Any cut-off of the higher frequencies produces, at the receiver, a blurred, out-of-focus effect, while interference produces bands, lines, or spots of varying shapes and arrangements, according to the source and nature of the interference.

Referring to Fig. 100, it will be seen that there are two unusual features in the "A" amplifier circuit. In the first place the photoelectric cells are connected directly to the grid of the first valve, no grid condenser being used, as is usually the case with talking picture amplifiers. The condenser is eliminated in this case because it would tend to cut off the higher frequencies. The second unusual feature is that the *positive* terminal of the high-tension photoelectric cell battery is earthed. The circuit as a whole is similar to the usual resistance-capacity coupled amplifier with the exception of the addition of the inductance L_2 in the second stage, and the tuned circuit L_1C in the fourth stage. The values of these components are so chosen that an increased gain results in the region of the higher frequencies. It is claimed that the amplifier gives a straight line response over a range of frequencies extending up to 18,000 cycles.

In Fig. 100 all neutralizing devices and grid bias arrangements have been omitted. The H.T. supply is 200 volts from a common battery, and the filament supply is 6 volts. The purpose of the switch labelled "phase switch" is to reverse the connections between the "A" and AX2 amplifiers, so that if it is found that a negative image is being broadcast, it can be instantly changed to a positive image by simply changing over the switch.

A negative image is so-called because it looks exactly like a photographic negative. It is impossible to be certain whether a positive or negative image is being broadcast until the image

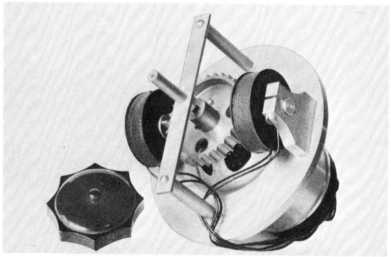

[*Photos Courtesy: Baird Television Ltd.*

Above : Component parts of the Baird synchronising gear, magnet coils, toothed wheel, and universal motor. *Below :* The assembled driving and synchronising unit.

[*To face page* 174.

PRESENT STATE IN ENGLAND 175

is tuned in on the check receiver shown in Fig. 99. Day to day changes in land-line connections through Savoy Hill to Brookmans Park, which do not affect sound broadcasting, cause this uncertainty. The transmission standard adopted is such that a positive image will be received by anyone using after the anode-bend detector of his wireless receiver a three-stage R.C. coupled L.F. amplifier.

The spot-light projector resembles an ordinary optical lantern, with the exception that a revolving scanning disc, about a foot in diameter, takes the place of a lantern slide. A 900-watt metal filament lamp is used in the lantern. The motor which drives the disc revolves at 750 r.p.m., which gives a scanning speed of $12\frac{1}{2}$ pictures per second. The motor is designed specially to run at a constant speed, and is loaded with a heavy flywheel. Even the current supply is constant, being supplied by a 12-volt battery.

A wireless receiver feeding into a check "televisor" completes the equipment of the control room. By watching the image in the check "televisor" the control engineer can rectify such faults in the image as may lie under his control.

A photograph of the Baird Televisor Receiver is shown facing page 168. It consists essentially of an electric motor for driving the scanning disc, the disc itself, a neon lamp of the flat plate type, a synchronising mechanism, and a lens system for magnifying the size of the image. The motor is of the universal type, i.e. it will run off either A.C. or D.C. mains. The disc, known as the Baird Graduated Disc, is shown in Fig. 101, which gives all the dimensions with the exception of the number of holes and the size of them. There are thirty holes, separated by a radial arc of 12°, and with the exception of the three holes at the beginning and end of the spiral, they are one-thirtieth of an inch square. The six end holes are oblong in shape, the object being to broaden the received picture by adding to each side three strips which are twice as wide as the central strips. This manœuvre lessens the detail at the sides of the picture, of course, but since there is never much of importance to be seen at the sides, this does not much matter. The advantage of the arrangement is that it is possible to improve slightly the detail of the central and important part of the image without increasing the signal frequency, thus making the utmost use of the available 9 Kc. broadcast frequency band.

The synchronising gear is of great interest, because it operates automatically, using the picture impulses themselves to keep the receiving motor in step, instead of requiring the local generation of control frequencies, or the transmission of such frequencies over a separate channel from the transmitter.

The device is called by Baird a cog wheel synchroniser, and he has gone out of his way to provide a most ingenious explanation of its action, so that it shall not appear to be just simply what it most closely resembles, a phonic wheel, or impulse motor, which, as we have already seen, will control the speed of a motor to which it is attached.

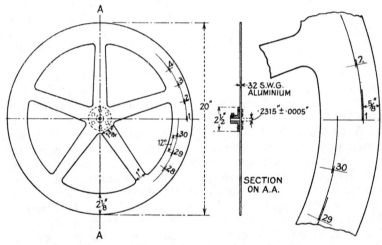

FIG. 101.—The Baird graduated scanning disc.

The explanation of the phonic, or cog wheel synchroniser commences at the transmitter, where the scanning beam, as it issues from the lantern, is masked so that light can only shine forth on to the sitter's face through one hole of the scanning disc at a time. This mask can be adjusted so that just as the light-spot from one hole is cut off, light is admitted through the next hole in succession. In that way there is at all times a light-spot on the face of the sitter from some hole or other. Or the mask can be adjusted so that there is a short interval of time between the masking of one spot and the unmasking of the next. During that interval there is no light on the sitter's face at all, no current flows in the photoelectric cell

PRESENT STATE IN ENGLAND

circuit, there is a definite break in the image signals, and each image strip at the receiver is slightly shortened, or in other words, there is formed a narrow black (dark) band across the top of the image. This band is, of course, invisible, so that the practical result is a slight shortening of the length of the image.

If the first method of masking is adopted, the image signal flows continuously, though modulated by the lights and shadows of the scene, i.e. the picture strips are all placed end to end in a long line without a break, as it were.

If the second method is adopted, the image signal flows intermittently, as well as being modulated; the picture strips, though placed end to end in a long line, have a definite break between the end of one strip and the beginning of the next. There are thirty strips, so that the frequency of the interruptions is thirty times $12\frac{1}{2}$ (the number of pictures sent per second), or 375 per second. Looked at in another way, 375 picture strips are sent per second. In a sense it is a combination of the interruption and modulation methods of signalling, and the practical result at the receiving end is a current which has two main components, the image frequency and a lower, clearly defined strip frequency of 375. In the "televisor" this strip frequency is utilised to actuate the phonic wheel and preserve synchronism, while the image frequency is, in a sense, filtered out and caused to operate the neon lamp.

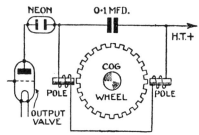

Fig. 102.—Connections of the Baird cog wheel synchroniser.

If no mask is provided at the receiver, several images can be seen at once, only one of which will be central in the viewing lens. Between each image the black synchronising strip can be clearly seen.

The connections of the cog wheel synchroniser are shown in Fig. 102. The cog wheel itself is of laminated soft iron, and has thirty teeth. The width of these teeth, and the width of the faces of the magnet pole pieces, is one quarter the width of the space between the teeth. Baird lays great stress on this fact, pointing out that in a phonic wheel the width of the teeth is the same as the distance between them. He also states that a phonic wheel will not give successful results. There is no doubt that it would, providing the dark strip between successive images was made wider—an undesirable move, and one which Baird avoids because his cog wheel variant of the phonic wheel gives him a patentable system of automatic synchronism. The phonic wheel, being long known to the art, is not patentable.

The cog wheel is mounted on the opposite end of the motor shaft to that on which the disc is mounted. The two electro-magnets are mounted on a ring which, through a geared knob, can be rotated about the cog wheel to facilitate correct image framing. The cog wheel device enables isochronism to be achieved, while the rotation of the magnets enables synchronism to be achieved.

Referring to Fig. 102, the output of the L.F. amplifier of the wireless receiver is fed first to the neon tube and then to the coils of the electromagnets, but these coils are shunted by ·1 mfd., a condenser which by-passes the high-frequency image currents, so that the current received by the coils is principally the 375 cycle strip frequency.

According to the explanation furnished by Baird, the action of the device is as follows:—

When the cog wheel is turned so that two teeth are exactly opposite the two pole faces, the transmitter scanning disc is in such a position that all holes are masked; that is to say, the active hole in the receiver disc has entered the dark strip, and no current is flowing, either through the coils or through the neon lamp. As the two teeth begin to leave the pole faces, the strip frequency current gradually builds up to full strength as the hole in the transmitter scanning disc begins to be opened to the light. The magnet poles are then energised and exert an attraction on the next two diametrically opposite teeth of the cog wheel. As these teeth come opposite the pole faces, the attraction ceases with the cessation of current.

Thus, as a tooth approaches a pole face, it is hastened by the attractive force, and as it leaves it, it is retarded by the next attractive force, until the next tooth comes within the range of attraction. These attracting and retarding forces, however, tend to cancel out for each cycle of changes, and the net result is that small overall changes of speed are opposed, and the motor to which the cog wheel is attached is "held" at a constant speed.

PRESENT STATE IN ENGLAND 179

If the motor should tend to run a little too fast, the result is that the period of no signal, instead of occurring when the tooth is dead opposite the pole face, occurs after it has passed it. But once the tooth starts to pass the pole piece while it is energised, a strong retarding action immediately takes effect and tends to pull it back. At synchronous speed this retarding action takes place later in the cycle. The retarding action thus predominates, and slows up the motor. If the motor should tend to run too slow, a converse action accelerates it.

However, the power available from the amplifier is decidedly limited, and in order to make the utmost use of it, it is desirable that the motor should be fitted with ball bearings, so that small speed variations can easily be overcome by the available correcting power without wasting energy in overcoming friction. Furthermore, the clearance between the pole faces and the teeth of the cog wheel must be very fine, and this necessitates very accurate machining of parts, which adds to the expense. Alternatively, an extra stage of amplification can be inserted in one lead to the magnet coils. This gives more power for synchronising, but again adds to the expense.

On the whole, this system of synchronising works very well indeed, provided that incoming signals are sufficiently strong and are not marred by interference or fading. It would be interesting to see the system applied to mechanisms which had been machined and balanced to the highest degree of accuracy obtainable, and including a ball-bearing motor.

At one time the transmitter driving motor used by the Baird Company was allowed to run freely, but now it also is governed by a cog wheel device which is supplied with 375 cycle current from a tuning-fork. This ensures that the transmitter motor shall at all times run at constant speed, and relieves the receiver synchronising mechanism of the necessity for following transmitter speed changes as well as correcting speed variations of its own motor.

Prior to the development of this system of synchronising, an automatic device known as the relay synchroniser was used. Although it has been superseded by the cog wheel synchroniser, it is of sufficient interest to describe here. The circuit details are shown in Fig. 103, from which it will be seen that there is a fixed resistance R_1 and a variable resistance R_2 in the field coil of the shunt-wound driving motor. Across the fixed resistance are connected the contacts of the relay, so that when current passes through the energising coil of the relay the contacts close and the fixed resistance becomes short-circuited, thus immediately increasing the field current, which slows up the speed of the motor.

The coils of the relay are in series with the neon tube in the output circuit of the last stage of the L.F. amplifier, but are shunted by two phosphor-bronze brushes which bear on a 30-segment commutator mounted on the motor shaft. Now, so long as the receiver motor is in step with the transmitter motor, both brushes bear on the same commutator segment during the whole length of an image strip. At the moment when one brush is on one segment and the other brush is on another segment, the short circuit is removed from the relay

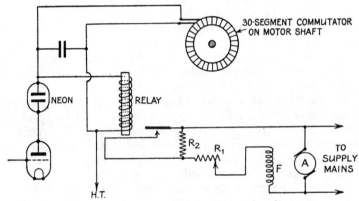

Fig. 103.—Details of the Baird relay synchroniser.

coils, and the relay would function but for the fact that at that instant the dark band of the image is reached, and there is a momentary cessation of current in the plate circuit of the last valve. But if for some reason the receiving motor gets out of step, then the transition of the brushes from one segment to another will cause a removal of the relay coil short-circuit at a time when current is flowing. The relay then functions, short-circuiting R_2, and reducing the speed of the motor.

The disadvantages of this method are that, in the first place, it is necessary to run the receiver motor slightly faster than the transmitter motor, because the device can only brake; it cannot accelerate. In the second place, there are mechanical objections to be found in the brushes and relay contacts, which

[*Courtesy: Baird Television Ltd.*

Close-up of Baird's large screen, employing a mosaic of 2100 tiny lamps.

[*To face page* 180.

PRESENT STATE IN ENGLAND

wear rapidly and require frequent cleaning and adjustment. The effect of these disadvantages on the received image was that the image continually swayed up and down by quite a considerable amount; it scarcely ever remained steady, and then only for a second or two, and the continual operation of the relay produced a chattering noise which was irritating to listen to for any length of time.

One of Baird's latest developments is a large viewing screen suitable for the exhibition of television images to a large audience. It made its bow to the public in August, 1930, when it was a daily feature on the programme of the London Coliseum for two weeks. Another development, also publicly exhibited at the same time and place, is called by Baird " tele-talkies," and consists in the transmission and reception of a talking film by television methods.

The large screen measures approximately 2 feet wide by 5 feet high, and consists of a mosaic of 2100 tiny flash lamp bulbs a few inches in front of which is placed a ground glass screen, the purpose of which is to diffuse the light from the bulbs. Each of the bulbs is connected by a single wire to a huge commutator containing 2100 segments. The return lead from each bulb goes to the frame, as shown in Fig. 104. Although the individual commutator segments are only $\frac{1}{32}$-inch in width, the diameter of the commutator is nearly 3 feet. A light arm, or selector switch mounted on the shaft of an electric motor, sweeps across the segments. One lead from the output of the image signal amplifier is connected to this arm, so that as it revolves it switches the television signal currents to each lamp in turn.

At the rate of transmission of $12\frac{1}{2}$ pictures per second, it is clear that each individual lamp is supplied with signal current for only $\frac{1}{12\frac{1}{2} \times 2100} = \frac{1}{26250}$th of a second. Although the metal filamented lamps normally take a current of only ·2 amp., it is clear that in order to bring them up to full brilliancy in such a brief space af time, they will have to be supplied with a much higher current. Actually, a small steady current is supplied which keeps the lamps just barely glowing, so that the signal current has only to increase the brilliancy from dull red to full brilliancy. Even so, it has been found necessary to supply a current of 5 amps. in order to produce the desired effect within the time available. This current is obtained by amplifying the signal impulses in a high-power amplifier,

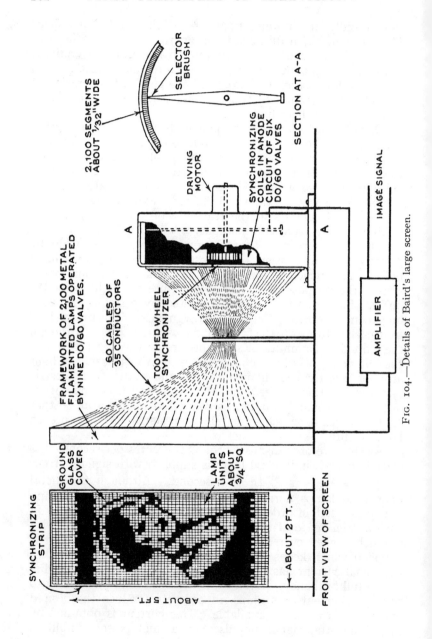

Fig. 104.—Details of Baird's large screen.

the output stage of which consists of nine DO/60 valves in parallel.

Synchronising is achieved by means of a toothed wheel synchroniser similar to that used in the smaller " televisor " but, since it has to control a much larger motor, part of the incoming signal is separately amplified by an amplifier having six DO/60 valves in parallel in the output stage.

The image produced on this screen is very bright, but is unrecognisable at close range. At a range of about 150 feet the image is quite recognisable, but is not equal in quality to that obtained on the smaller " televisor " ; it is not so rich in half-tone effects.

The transmission of talking films is very simply effected by running the film through a projector and focusing the picture

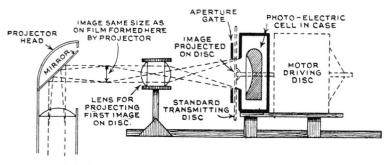

FIG. 105.—Baird's " tele-talkie " arrangement for transmitting talking films.

on to the scanning disc behind which is placed a single photo-electric cell, as shown in Fig. 105. The usual type of intermittent projector cannot be used, however, as special mechanism would be required to adapt it from the speed of 22 pictures per second for which it is designed, to $12\frac{1}{2}$ pictures per second, as used by Baird at present. There is a type of projector available in which the film runs continuously and smoothly, an optical arrangement of mirrors and lenses causing the pictures to be superimposed into each other. This type of projector produces no flicker, no matter how slowly it is run. The objection to it for normal cinema use is that too much light is lost in its complicated optical system, but for the transmission of films by television methods it is ideal.

The sound part of the transmission is recorded on the edge of the film as a ladder-like structure, and is reproduced during transmission in the usual manner, which is to focus an intense pin-point of light which pierces the sound track, the different densities of which vary the intensity of the emerging light, which then falls on to a photoelectric cell, as shown in Fig. 106. The output of the cell, fluctuating in accordance with the sounds recorded on the film, is then amplified, transmitted by wire or wireless, and reproduced on a loud speaker in the auditorium, just as is done in the case of a talking film in a cinema.

It is not within the scope of this book to go into details concerning the details of the wireless receiver to be used to pick up television broadcasts. Suffice to say that two receivers are required at present, one to pick up the sound part of the programme in the usual manner, and a second to receive

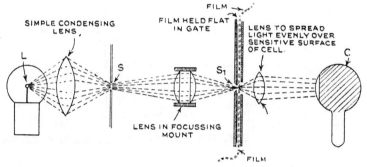

Fig. 106.—Diagram of a talking picture sound-reproducing unit, operating from the sound-on-film method of sound recording.

the television signals from the second transmitting station. The main requirements for the latter receiver are that it shall be capable of delivering a strong clear signal, and its output must be undistorted. That is to say, it must be capable of reproducing at equal strength signals occurring at any frequency within the audio frequency band, to which television signals are at present confined.

In operating the "televisor," several phenomena may present themselves. On first switching on, a kaleidoscopic collection of bright streaks, dots and bars will be seen as the motor runs up to speed. Gradually the image will take shape and appear to be flying round at a great rate past the viewing lens. Bit by bit the image will slow down and come to rest, if the speed control is properly adjusted. In all probability it will now be found that the image looks like Fig. 67—split vertically down the centre. This is due to the fact that the synchronising gear has done its work at an unpropitious moment and is holding the receiving disc out of phase with the transmitting disc. In other words, the holes do not match up. This is very easily remedied by turning the speed control slightly and gently one way or the other until the image begins to drift round again. When it is seen to be getting near the correct position, gently turn the knob back again, and the image will be held in its proper position. The knack of making this adjustment is soon acquired. The motto is : Do It Gently. Any violent movement of the speed control will send the image flying round at a speed too fast to follow, and it will be some time before control can be regained.

If the image is seen as in Fig. 68 it is improperly framed, but can easily be correctly framed by turning the framing knob which controls the position of the synchronising coils by swinging them round the periphery of the phonic wheel.

It may be that a negative image will be received. If this should happen to be the fault of the transmitting station it will very quickly be corrected. If not, then the trouble will have to be rectified in the wireless receiver. If a resistance-capacity coupled amplifier is in use it will be necessary to add another stage in order to reverse the current phase, but if a transformer coupled amplifier is being used it will suffice to reverse the connections to one winding of one of the transformers. In either case, if an output transformer is in use (instead of putting the neon tube direct into the plate circuit of the last valve), it will be sufficient to reverse the connections to the primary.

Interference will produce patterns on the "televisor" screen which will vary according to the cause of the interference. Induction from local A.C. mains will produce wavy lines across the picture. Strong atmospherics will produce a snowstorm effect, while the music of an interfering broadcasting station will be apparent as an interweaving, endlessly varying pattern of not inconsiderable beauty.

When receiving from a distant station at night the phenomenon known as echo images may be expected to manifest itself. When receiving sound from a distant station there are times when the timbre of the sound alters and becomes somewhat hollow. Following this there may be a period of fading.

When receiving television images during the prevalence of such conditions, a second "ghost" image will be observed which, if vertical scanning and an anti-clockwise running disc is used, as in the Baird "televisor," will be slightly above the main image. This ghost image gradually becomes brighter, while the main image fades. Sometimes the main image will disappear altogether, leaving a full-strength ghost image set too high in the frame, until after a few seconds the main image will reappear and the ghost will disappear. Or both images may disappear if fading is bad. Sometimes three and four images can be seen at once, one above the other, the space between images varying with conditions.

The reason for this effect is that at night there are two kinds of wireless waves arriving at the receiver, the ground wave and the space wave. The ground wave, as its name implies, follows the surface of the earth from transmitter to receiver and provides the most reliable signal, especially during the day. The space wave shoots off from the transmitter at a tangent to the surface of the earth until it reaches the ionised Heaviside Layer, when it is reflected back down again to earth. This wave, having a longer path to follow, arrives at the receiver a fraction of a second later than the ground wave, producing a slight echo effect in the case of sound reception which, however, is so slight as to be rarely perceptible to the average listener.

With television reception, however, the case is different. The speed in transmission is so great, and our eyes so much more efficient than our ears, that the delayed wave produces a second set of signals which build up a second image which our eyes can readily detect. If both waves arrive in phase, and are of equal strength, the two images are equally brilliant; otherwise the stronger wave produces the brighter image. If, on the other hand, the space wave arrives 180° out of phase with the ground wave, and is of equal strength, the two cancel out and no signal results, i.e. there is a complete fade-out. If the out-of-phase space wave predominates, then a negative ghost-image results which will remain until the ground wave again predominates, or until the space wave reverses its phase. Three and four images are produced by several space waves arriving by different routes, after having been reflected from the Heaviside Layer at different points. All sorts of complicated combinations of these effects may produce all sorts of multiple image effects, and the phenomenon constitutes, at

[*Courtesy: Jenkins Television Corporation.*

The Jenkins radiovisor in its simplest form, using a disc scanner, together with special Jenkins wireless receiver for television signals.

[*To face page* 186.

PRESENT STATE IN ENGLAND

present, one of the drawbacks to the regular employment of wireless as a means of transmitting television over great distances.

There are one or two other reception troubles which are purely local and under the control of the operator. If blank dark patches appear in parts of the images, and persist, there is not sufficient bias voltage on the neon tube. If, on the other hand, the picture is being swamped by too much brilliance, reduce the bias voltage. A faint, scarcely discernible image indicates that there is insufficient amplification. If a volume control is fitted to the amplifier, raise the volume. If the image appears with too harsh a contrast, entirely lacking in half-tone effects, the amplification is too great and the picture is being "over-cooked." Use the volume control to reduce the degree of amplification.

If, after checking over all these adjustments, it is found that the person in the image has a halo round his head and spurious shadows under the eyes, nose and chin, the trouble is due to the fact that the amplifier and/or wireless receiver in use is cutting off the lower frequencies. If, on the other hand, there are no spurious shadows, but the image appears flat and lacking in detail, the receiver and amplifier combination is cutting off the higher frequencies. The receiver-amplifier combination may appear to function perfectly on music reception, but, as already stated, the eye is a more efficient fault-finder than the ear.

Quite a large number of amateur experimenters are being attracted to television in England to-day, largely because television represents something new in the wireless field, but at the time of writing the non-technical public is not displaying sufficient interest to make television a commercial proposition. The reason is that television to-day may be said to be in the same state as wireless was before the war; there were a number of wireless enthusiasts in those days whose interest was purely technical, but there was nothing for the non-technical man to listen to except Morse code, which he couldn't understand. Television to-day is very interesting to the technical man, but it can show the non-technical man but little, and once he has seen it two or three times, there is nothing in the present crude and limited images to hold his interest once the novelty has worn off.

The introduction of sound broadcasting added the touch of sustained interest to wireless which made the non-technical man, after hearing one or two programmes on his technical neighbour's set, go round to the nearest wireless dealer and order a set of his own. Television has not yet reached that point, and until it does it will remain a nine days' wonder to the man in the street, an intriguing novelty to be seen once as one of the " sights," and forgotten about until some new development crops up. That this should be the case is but natural to an infant art, and will readily be recognised and acknowledged by most people. But nothing but harm can be done to the struggling infant if the facts are overlooked and a deliberate effort is made to overstate the claims of television in its present state, and commercialise it prematurely. And it is definitely too early to attempt to commercialise (on anything but a limited scale among amateur enthusiasts) any of the systems of television which are being experimented with to-day.

189

CHAPTER XIV.

THE PRESENT STATE OF THE ART IN AMERICA.

MECHANICAL SYSTEMS.

Jenkins Radiovisors and Broadcasts. Bell System Two-Way Television. Method of Synchronising. The Acoustic System. Alexanderson's Large Screen. Appendix—Wavelength and Power Assignments for Experimental Television Broadcasts in America.

AT the time of writing, there are in the United States five leading workers in the television field. These are the Jenkins Television Corporation, the Bell Telephone Laboratories, Dr. E. F. W. Alexanderson, of the General Electric Company, Dr. V. Zworykin, late of the Westinghouse Electric and Manufacturing Company, and now associated with the R.C.A. Victor Company, and Philo T. Farnsworth, of the Television Laboratories, Inc., San Francisco.

The Jenkins Television Corporation has taken over the patents and inventions of C. Francis Jenkins, and is endeavouring to popularise television by placing on the market simple and relatively inexpensive apparatus, and by broadcasting television regularly. These broadcasts are sent out daily, between 2 and 4 p.m., and between 5 and 7 p.m. E.S.T. (7 to 9 p.m. and 10 to 12 p.m. G.M.T.), from station W2XCR, Jersey City, N.J., on a wavelength of 147·5 metres (2035 Kc.). The accompanying speech is transmitted on 254 metres by WGBS, one of New York's broadcasting stations, in the studio of which the Jenkins transmitter is installed. The transmissions include ordinary talking films showing the head and shoulders of the subject only, and also direct television pick-ups from human subjects. In passing, it may be mentioned that details of all American stations authorised to broadcast television images are given in an appendix at the end of this chapter.

The Jenkins Corporation is marketing two types of receiver. both of which are shown in the illustrations facing pages 186 and

194. The simplest type, which sells for £15, consists simply of a motor, scanning disc, magnifying lens, and toothed wheel synchronising mechanism almost exactly similar to that used by Baird.

The Jenkins picture is square in outline, is composed of 60 scanning lines, horizontal scanning is employed, and the speed of transmission is 20 pictures per second. The strip or synchronising frequency is therefore 1200, and the picture frequency is 36,000 cycles. At the receiver the 1200 cycle synchronising frequency is filtered out from the incoming signal, fed to a special amplifier, and thence to the synchronising coils, which operate on the toothed wheel to preserve synchronism in exactly the same manner as in the case of the Baird synchroniser.

It will be observed that the short wave receiver for the Jenkins signals must be capable of receiving and amplifying without distortion a 36 Kc. frequency band. Very few receivers are capable of doing this, so the Jenkins Corporation also market a special wireless receiver, complete with all-mains equipment, power amplifier, and valves for £40. At their transmitting station the Jenkins Corporation have developed an image signal amplifier capable of straight line amplification over a range of frequencies from 15 to 50,000 cycles.

The larger model radiovisor is a development of the radio-movie receiver described in Chapter V., and is marketed at £80. Instead of being mounted horizontally, as formerly, the motor and drum scanner are now mounted vertically. The multiple target neon lamp has been replaced by a single flat plate neon lamp, and the quartz rods and selector switch have been eliminated. In place of the selector switch there is now a shutter consisting, in the case of a receiver designed for 48 line scanning, of a disc with four curved slots, which rotates in front of the scanning drum. The scanning drum makes four revolutions for each complete picture of forty-eight lines, and the selector shutter serves to mask all but the twelve holes which are active during each revolution. The shutter is driven at one-quarter the speed of the drum, through reduction gearing. The general appearance of the receiver (with case removed) is sketched in Fig. 107. The external appearance of the receiver, mounted in its case, can be seen from the illustration facing page 194.

A study of the terms under which television broadcasting licences are issued (see appendix to this chapter) will reveal that the Jenkins Corporation, like everybody else in America,

PRESENT STATE IN AMERICA

at the time of writing, is limited to experimental transmissions only, and may not commercialise these transmissions, i.e. sell advertising "time on the air." The possible market for radiovisors, therefore, is rather limited to keenly interested amateurs at the present time.

The Bell Telephone Laboratories, Inc., demonstrated during 1930 their latest development, two-way television by wire as an adjunct to the telephone. For the purposes of this demon-

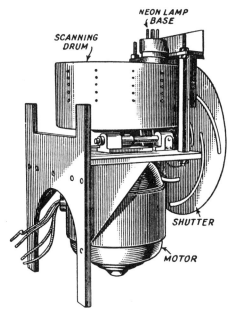

FIG. 107.—The chassis of the improved Jenkins radiovisor.

stration, experimental equipment was set up in the headquarters building of the American Telephone and Telegraph Company at 195 Broadway, and at the Bell Telephone Laboratories at 463 West Street, New York, the distance between these two buildings being about 2 miles. The object of this experimental system is, of course, to enable two persons holding a telephone conversation also to see each other, and the terminal equipment, at each end of the circuit, is placed in a special telephone booth. A schematic diagram of the layout of the system is given in Fig. 108.

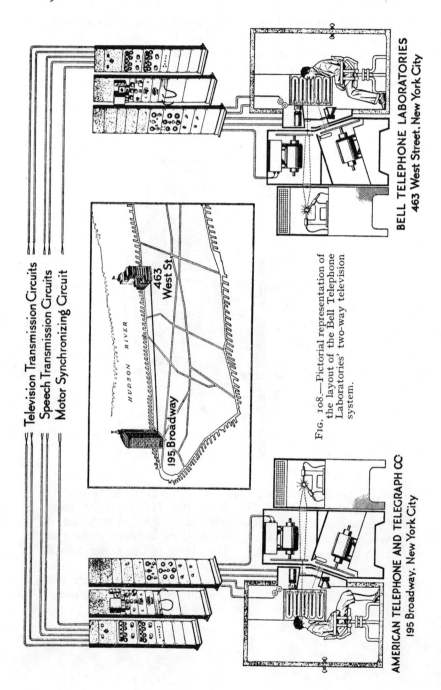

Fig. 108.—Pictorial representation of the layout of the Bell Telephone Laboratories' two-way television system.

PRESENT STATE IN AMERICA

The system consists, in essence, of a duplicated disc arrangement, with two scanning discs, one for transmission and one for reception, together with a neon lamp and bank of photoelectric cells at each end of the circuit. In what follows, the terminal equipment at one end only will be described, that at the other end being an exact duplicate.

The transmitting disc (top disc in Fig. 108) is 21 inches in diameter. Three banks of photoelectric cells, making twelve in all, are arranged on either side and above the person's face and serve to pick up the reflected light and generate the television signals. The second disc, which is 30 inches in diameter, is placed below the transmitting disc, and inclined slightly at an angle to make it more convenient for the telephone user to watch the glowing plate of the neon tube through the holes of the rotating disc.

A problem of illumination is immediately encountered in this system, which is that any dazzling form of spot-light makes it difficult or impossible for the telephone subscriber to see the dim (by contrast) glow of the neon lamp. This difficulty has already been commented upon in describing a similar demonstration equipment which was shown at the 1929 Berlin Radio Exhibition. In the Bell system, this difficulty was overcome by interposing a blue filter in the scanning beam. The human eye is relatively insensitive to the consequent blue light which falls on the face. As we have already seen, photoelectric cells of the potassium type are more sensitive to light in this region of the spectrum, and the potassium cells used in the present system are specially sensitised with sulphur for the particular quality of blue light employed. In contrast to this, the telephone booth is illuminated by an orange-coloured lamp, to the rays of which the cells are insensitive. The number and size (active area) of the photoelectric cells has also been increased (by comparison with those used in the 1927 demonstration) so that the intensity of the scanning beam could be reduced.

Although easily recognisable images could be obtained with the 50-hole discs previously used, it was decided for the two-way demonstrations to increase the number of holes to 72, so as to improve the images to a point where individual facial characteristics could be identified without the slightest difficulty. This increase in the number of holes exactly doubles the number of picture elements in the image, and increases the frequency of the picture signals from 20,000 to 40,000 cycles. The number of images transmitted per second remains the same, that is, 18.

In order to obtain still more brilliancy in the received image, and help to overcome any remaining dazzling effect from the blue scanning beam, a water-cooled neon tube was at first employed which enabled more power to be delivered to the tube. The design of these tubes can be seen from the photograph facing page 210. Heavy metal bands attach the rectangular cathode to a hollow glass stem occupying the central portion of the tube. Water from a small circulating pump flows continuously through the glass stem and cools the cathode by thermal conduction through the metal bands. To reduce sputtering of the electrode and consequent blackening of the glass walls, the front surface of the cathode is coated with beryllium. This metal resists the disintegrating action of the glow discharge very satisfactorily and gives the lamp a prolonged life. Other metal surfaces of the tube are shielded from the discharge by mica plates. The increased brilliancy of the tube is thus obtained by overloading it.

The water-cooled plate type tubes have now been superseded by large air-cooled crater type neons which are used in conjunction with a colimating lens placed between the tube and the back (from the point of view of the looker-in) of a lens disc.

Pure neon in a plate or crater type of lamp gives a very inferior reproduction of the image. The impedance of the lamp is relatively high and comprises both a resistance and a reactance which vary with frequency. The variation in the impedance causes a relative loss in the frequency components of the signal, and also introduces spurious phase shifts. In addition, pure neon has an afterglow, the gas continuing to glow for an appreciable time after the current ceases to flow. This afterglow casts spurious bands of illumination out to one side of the brighter image details. A small amount of hydrogen in the neon prevents such an afterglow, and at the same time improves the circuit characteristics of the lamp. The total impedance of the lamp is lower, making it a less influential part of the lamp circuit ; and the resistance and reactance vary in such a manner that the phase shift is more nearly proportional to frequency. A phase shift proportional to frequency causes no distortion in the reproduction of an image. Other active gases may be used with the neon to improve the operation of a television lamp, but one or two per cent of hydrogen is most satisfactory.

[*Courtesy: Jenkins Television Corporation.*

A console model Jenkins radiovisor, using a drum scanner (right). A special wireless receiver is also incorporated (left).

[*To face page* 194.

PRESENT STATE IN AMERICA

Since hydrogen is absorbed by the electrodes during a glow discharge, it slowly disappears from the neon during operation of the lamp. For this reason the lamp is provided with a small side reservoir of hydrogen. Even with this improvement it is admitted that the circuit characteristics of the lamp are not ideal.

The overall size of the plate type of lamp is about 9 inches, and that of the crater type about 12 inches long by 6 inches in diameter.

To enable the technical operator to monitor the outgoing television signals a small neon lamp is placed behind the transmitting disc but displaced several frames from the aperture through which the arc beam passes. By continuing the spiral of holes part way round it is possible to see the complete image by means of the auxiliary neon lamp, to which the outgoing signals are also applied. In order that this may be seen from the operator's position behind the disc, a right-angled prism and a magnifying lens are placed in front of the disc, and the image is seen through an opening in the side of the motor cabinet. The task of the operator is to direct the scanning beam up or down by means of a variable angle prism until the face of the person in the booth is centrally positioned. Having found this position for each individual, according to his height, the operator's next task is to raise or lower the large magnifying lens which is placed in front of the main receiving disc and neon lamp, so that the subscriber seated in the booth can comfortably and properly see the received image. Optical arrangements are also provided to enable the operator to view the main incoming image, and see if all is well. If any adjustments are required he can make them and watch their effect upon the image.

The synchronising methods used in this equipment are bound up in a different type of driving motor to that employed during the 1927 demonstrations, together with simpler and cheaper regulating apparatus. The motor is a 4-pole compound wound D.C. motor with the following special features added:

(1) An auxiliary regulating field, the current through which is controlled by a thermionic valve regulator.

(2) A damping winding on the face of the field poles to prevent the field flux from shifting (see Fig. 109).

(3) Three slip rings are provided at points 120 electrical degrees apart for furnishing three-phase power to supply plate and filament voltage for the regulating circuit.

(4) A pilot generator of the inductor type is built into the motor frame and delivers a frequency proportional to the motor speed for actuating the control circuit.

(5) A hydraulically damped coupling is provided between the motor shaft and the scanning disc (see Fig. 110).

Features 2 and 5 are adopted to prevent hunting or instability of the image. The hydraulically damped coupling, the construction of which is shown in Fig. 110, employs a flexible metal bellows filled with oil and connected by a small pipe equipped with a needle valve for adjusting the amount of damping. The scanning disc itself is centred on a ball bearing which allows the

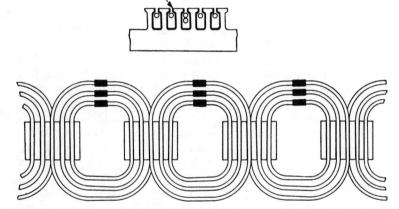

Fig. 109.—Damping winding of closed turns which prevents shifting of field flux.

disc to rotate with respect to the shaft within approximately plus or minus 5° mechanical movement.

The following description of the equipment and its operation, together with the diagrams, are taken from the *Bell System Technical Journal*, Vol. IX., No. 3.

Referring to Fig. 111, when the motor is operating at full speed the pilot generator delivers approximately 1 watt of power at 300 volts, 1275 cycles to the plates of a pair of push-pull detector valves. The grids of these valves are supplied with an E.M.F. of the same frequency from an oscillator or other source of power having a sufficiently constant frequency. The amount of power required for this grid circuit is only a few

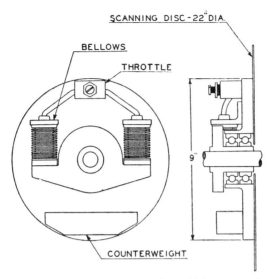

Fig. 110.—Hydraulically damped coupling which prevents hunting of the motor.

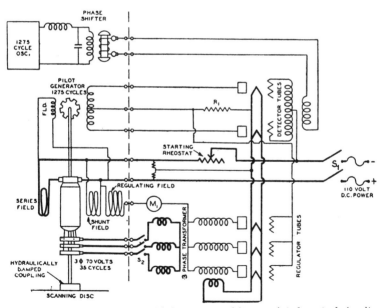

Fig. 111.—Schematic diagram of the motor and its associated control circuits.

thousandths of a watt. The detector valves rectify the plate voltage, producing a potential drop across the coupling resistance R_1. If the plate and grid voltages are in phase, so that the grids of the valves are positive at the same instant that the plates are positive, the plate current will be a maximum. If the grid voltage is negative when the plate voltage is positive, the plate current is practically zero, so that the magnitude of this current is a function of the phase relationship between the grid and plate voltages.

The voltage drop across the coupling resistance R_1 is applied to the grid circuits of three regulator valves. These valves derive their plate voltage supply from a 3-phase transformer fed with power from the three slip rings provided on the motor. The regulator valves act as a rectifier whose output is controlled by the potential impressed upon the grids from the coupling resistance R_1. The current of the regulator valves is passed through the regulating field provided on the motor. This field is in a direction to aid the shunt and series fields of the motor.

The operation of the circuit is as follows: In starting, switch S_1 is closed, which applies direct current to the shunt field and armature circuits of the motor. The motor accelerates as an ordinary compound-wound motor. Switch S_2 is then closed, applying three-phase power from the slip rings of the motor to the transformer. As the speed of the motor approaches the operating point, the beat frequency between the pilot generator and the oscillator will cause beats in the current through the regulating field which are visible on the meter M_1. Let us assume that the field rheostat has been previously adjusted so that with the shunt field alone the motor will tend to run slightly over the desired operating speed. When the exact operating speed is obtained, the beat frequency in the regulating field will be zero, and as the motor tends to accelerate, the phase relationship between the pilot generator and the oscillator will reach a point tending to give maximum strength to the regulating field. When this point is reached, the acceleration of the motor will be checked by the increased field and the speed will tend to fall until the phase of the pilot generator with respect to the oscillator has shifted sufficiently so that the regulating field current is reduced to an equilibrium value, after which the motor continues to run at constant speed in accordance with the frequency of the oscillator.

[*Courtesy: Bell Telephone Laboratories.*

The telephone booth used in the Bell system two-way television demonstrations. The received image is seen in the lower part of the central opening, while the scanning spot for transmission emerges from the opening above the image. The photoelectric cells can be seen through the glass panels on both sides and at the top.

Operating tests on the circuit show that the motor, when operating off 110 volts D.C., will hold in step over line voltage ranges from 100 to 125 volts, and will be self-synchronising over somewhat narrower voltage limits. Thus, under normal conditions all that is necessary from an operating standpoint is to close the switch and wait for the motor to pull into step.

The control oscillator is a standard type of thermionic valve oscillator having a frequency precision of the order of one part in 1000, when delivering the negligible output of ·005 watts to the grid circuit of the detector valves. This frequency is delivered directly to the motor circuits at one end of the line and is transmitted over a separate cable pair to the control circuits at the other end of the line. It was found that the detector valves would operate successfully over a considerable variation in power level, provided the minimum oscillator output was sufficient.

Framing of the image is very simply accomplished with this apparatus. Referring to Fig. 111 it will be noted that a phase shifter is provided between the oscillator and the input terminals to the control circuit. This phase shifter is designed with a split phase primary member producing a rotating magnetic field. The secondary member is single phase and is mounted on a shaft provided with a handle. By rotating the handle of the phase shifter in the desired direction, the frequency delivered from the phase shifter will be the algebraic sum of the frequencies of the oscillator plus the frequency of rotation of the armature of the phase shifter. It is, therefore, a simple matter for the operator at the receiving end to momentarily increase or decrease the control frequency and thus bring the picture into frame.

This system, although very complicated, operates with great success and reliability. As an example of its performance, if the television image is disturbed by a momentary load such as the pressure of a hand against the disc, it will come back to rest within approximately one second, there being two oscillations during this interval. In actual operation, it is found that the normal fluctuations in line voltage occurring on the commercial power supply produce no transients of sufficient magnitude to cause any objectionable instability in the received image.

A similar system of control could be applied to an A.C. motor by substituting a saturating reactor in place of the regulating field winding.

There remains only the acoustic portion of the system. Obviously, if an ordinary telephone set were employed, the subscriber's face would be obscured, so it has been eliminated. Instead, a sensitive condenser-type microphone, such as is used in broadcasting studios, is employed, together with a small loud speaker of the moving coil type with a small piston diaphragm. The microphone is placed in front of the telephone user as he sits in the booth in front of the television apparatus, and just above the aperture through which the scanning beam issues. The loud speaker is placed just below the magnifying lens through which the received image is viewed. Distortion, throughout the acoustic system, has been reduced to a minimum, so that the transmitted sounds shall be familiar and natural.

In any system arranged as described above, the microphone is unable to differentiate between the sound of the local human speaker and the sounds issuing from the adjacent loud speaker. If the sounds from the loud speaker are picked up by the microphone in sufficient magnitude, then " singing " will result, as in the case where an ordinary telephone receiver is placed in front of the mouthpiece. To prevent this, the microphone and loud speaker are installed in carefully chosen positions, and partly shielded from one another. Also the whole design of the booth has received much attention to the end that resonances shall be eliminated, and the walls, roof, and floor are treated with sound absorbing material. The only remaining sound reflectors of serious importance are the photoelectric cells, and these have been placed in such a way, and farther back than was the case during the 1927 demonstrations, that no serious reflection takes place from their glass walls.

Operating arrangements are so made that the two parties to the conversation, after taking their seats in the booths, do not see or hear one another until the technical operators have made all necessary adjustments. Then the operators expose the images and connect the talking circuits simultaneously. When the conversation comes to an end and the parties to it leave their seats in the booths, the act of getting up from the chair interrupts the connections.

The general effect of the whole combination on a party to such a conversation, and the degree of realism, may best be described by comparing it to holding a conversation with another person across a room at a distance of about 12 feet.

The improvement effected in the system by using 72-hole discs, together with the many other refinements, may be judged from the fact that the image is now so clear that when the test was made, expert lip readers were able to understand what was being said by the speaker at the other end of the wire when the speech circuit was switched off, simply by following the movements of the speaker's lips as seen in the television image. The scope of the image is, of course, limited to a head and shoulders view of the distant speaker.

Ever since 1927, Dr. E. F. W. Alexanderson, consulting engineer of the General Electric Company and the Radio Corporation of America, has been experimenting with television, using the disc system of scanning. His efforts have very largely been concentrated on the problem of enlarging the received image, and in May, 1930, his experiments culminated in a spectacular demonstration at a cinema theatre in Schenectady, N.Y.

On this occasion the artistes performed at the G.E.C. laboratories about 2 miles from the theatre, the television signals being transmitted by wireless on a wavelength of 140 metres, while the accompanying voice signals were broadcast on 92 metres. At the theatre two wireless receivers were installed, the output of the voice receiver being fed to two large loud speakers, while the television signal output of the other was used to control a light source to project the television image on to a screen measuring 6 feet by 7 feet.

The television receiver used for this demonstration was essentially similar to the 3-foot screen receiver used by Karolus and described in Chapter XII. The various elements of the receiver are illustrated in Fig. 112, while the complete details are shown in Fig. 113 and in the photographs which face page 222.

Referring to Fig. 112, it will be seen that the essential elements are an arc light, two crossed nicol prisms and a Kerr cell to control the intensity of the light beam, and a spiral lens disc with which to scan the screen.

As already explained, the principal disadvantage of the nicol-prism-Kerr-cell optical system is the great loss of light involved, and in the present system this difficulty has been overcome partly by brute force methods, and partly by improvement in the Kerr (or Karolus) cell. Dr. Alexanderson uses the largest moving picture projector available, the arc light of which consumes 150 amperes. He has, of course, retained only the projection part of the machine, and discarded the film reeling mechanism. In the original Karolus cell,

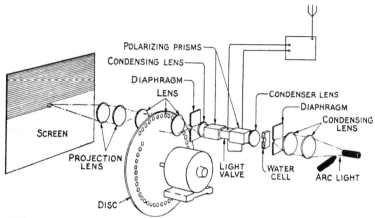

FIG. 112.—Skeleton drawing of the essentials of Dr. Alexanderson's large projection receiver.

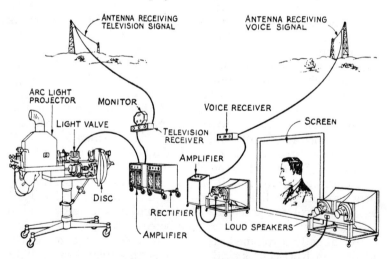

FIG. 113.—Pictorial representation of the complete layout of Dr. Alexanderson's large screen television receiver.

only two control electrodes were employed. Dr. Alexanderson succeeded in improving the cell considerably by using a number of electrodes, interdigitated like the plates of a condenser or accumulator. The polarised light beam issuing from the first nicol prism passes between these electrodes, the increase in the number and proximity of which gives a considerably augmented control over the light beam. It will be noticed in Fig. 112 that a water cell is included. The purpose of this is to provide water cooling to prevent injury to the nicol prisms or Karolus cell, due to the intense heat from the powerful arc nearby.

The whole of the apparatus shown in Fig. 113 is mounted on wheels so that it can be easily moved on and off the stage of a vaudeville theatre to make way for other " turns." The projector is placed at the back of the stage, 17 feet from the screen, which is semi-transparent to enable the audience on the other side of it to see the image. Synchronism of the motor which drives the lens disc is effected by means of a synchronous motor coupled to it. This unit can be seen in Fig. 113, and also in the photograph facing page 222, immediately below the Karolus cell, which in Figs. 112 and 113 is marked " light valve."

The images given by this system showed the head and shoulders only of the subject, and gave as much detail as is to be expected from a 48-line image. The degree of brilliancy of the image may be described as being about half that of a normal moving picture screen, but the image could be readily discerned from all parts of the theatre.

APPENDIX TO CHAPTER XIV.

Wavelength and Power Assignments for Experimental Television Broadcasts in America.

THE International Radio Convention of 1927, signed in Washington on the occasion of the last general International Radio Conference, did not provide specifically for frequency assignments to the new service called visual broadcasting, the most important subdivision of which is television.

A North American Conference, held in Ottawa, Canada, in January, 1929, set aside the following frequencies for television assignments:

        2000 to 2100 kilocycles
        2100 to 2200 ,,
        2750 to 2850 ,,
        2850 to 2950 ,,

with the additional frequency band 2200-2300 kilocycles, available for assignment in the United States, in such geographical regions as the South and South-West, where such assignments would not interfere with the use of the same frequencies for other purposes in Canada or any other nation on the North American continent, or in the West Indies. It will thus be seen that there are only four frequency bands, each 100 kilocycles wide, for general allocation in the United States.

The Federal Radio Commission (which controls broadcasting in the United States) has maintained a policy of permitting and encouraging the type of legitimate experimental research work in television which will advance the art. For this reason the present frequency assignments to television stations are made upon a purely experimental basis. All of these stations are subject to the provisions of General Order 64, covering experimental stations. This order requires the filing of regular quarterly reports showing the technical progress made by the station during the previous quarter, and definitely precludes the commercialising of the stations' transmissions.

The following list shows the television stations now licensed on an experimental basis by the Federal Radio Commission, together with the call letters, frequencies, and power used. As will be seen, there are eighteen companies engaged in television research work, operating twenty-two stations.

List of Stations Licensed to Transmit Television Experimentally in the United States, as at December 31st, 1930.

2000-2100 Kc.

Call Letters.	Company.	Location.	Power (watts).
W3XK	Jenkins Laboratories,	Wheaton, Md.	5000
W2XCR	Jenkins Television Corporation,	Jersey City, N.J.	5000
W2XAP	Jenkins Television Corporation,	Portable	250
W2XCD	DeForest Radio Company,	Passaic, N.J.	5000
W9XAO	Western Television Corporation,	Chicago, Ill.	500
W2XBU	[1] Harold E. Smith,	North Beacon, N.Y.	100
W1OXU	[2] Jenkins Laboratories,	aboard cabin monoplane	10
W1XY	Pilot Elec. and Mfg. Co.,	Springfield, Mass	250
W1XAE	Westinghouse Elec. and Mfg. Co.,	Springfield, Mass	20,000

2100-2200 Kc.

W3XAK	National Broadcasting Company,	Bound Brook, N.J.	5000
W3XAD	R.C.A. Victor Company,	Camden, N.J.	500
W2XBS	National Broadcasting Company,	New York	5000

[1] One hour daily (1 to 2 p.m. E.S.T.—6 to 7 p.m. G.M.T.).
[2] Daylight hours, week days only.

[*Courtesy: Bell Telephone Laboratories.*

Bell system two-way television equipment. At left is the arc projector; top centre, transmitting scanning disc; bottom centre, receiving disc. The tubes take the water supply to the water-cooled neon tube. At right are the backs of the photo-electric cells and first amplifier stage compartment.

| Call Letters. | Company. | Location. | Power (watts). |

W2XCW—General Electric Co., South Schenectady, N.Y. . 20,000
W8XAV—Westinghouse Elec. and Mfg. Co., East Pittsburgh, Pa. 20,000
W9XAP—Chicago Daily News, Chicago, Ill. . . . 1000
W2XR—[1] Radio Pictures Inc., Long Island City, N.Y.. . 500

2750-2850 Kc.

W2XBO—United Research Corporation, Long Island City, N.Y.. 500
W9XAA—Chicago Federation of Labour, Chicago, Ill. . . 1000
W9XG—Purdue University, West Lafayette, Ind. . . 1500
W2XBA—WAAM Inc., Newark, N.J. 500
W2XAB—Columbia Broadcasting System, New York . . 500

2850-2950 Kc.

W1XAV—Shortwave and Television Labs. Inc., Boston, Mass. 500
W2XR—Radio Pictures Inc., Long Island City, N.Y. . . 500
W9XR—Great Lakes Broadcasting Company, Downers Grove, Ill. 5000

The American Radio Manufacturers' Association has created a television committee, which, among other things, has considered and adopted standards of scanning as follows:
1. Scanning from left to right and from top to bottom.
2. A scanning speed of fifteen pictures per second.
3. The use of forty-eight lines, requiring forty-eight holes in spiral in the scanning disc.
4. In addition to the forty-eight-line standard, a secondary standard of sixty lines was created in order to permit more detail in a picture for more advanced research workers.

The committee does not require adherence to these standards. However, if one method of reception can be used for all television stations a larger experimental audience is obtained, since no receiving set experimenter has the inclination or the apparatus to shift from one to another of several methods of scanning, when he could be concentrating his efforts on reception experiments which would be independent of scanning methods.

[1] Subject to operation between 5 and 7 p.m. E.S.T. (10 to 12 p.m. G.M.T.). Subject to shared operation after 10 p.m. and before 2 p.m. E.S.T. (3 a.m. and 7 p.m. G.M.T.) by agreement with other licensees within 150 miles of W2XR.

CHAPTER XV.

THE PRESENT STATE OF THE ART IN AMERICA (*continued*).

Cathode Ray Systems.

Zworykin's Film Transmitter. Kinescope Receiver. Method of Synchronising Pros and Cons of Cathode Ray Method. Farnsworth's Dissector Tube. Electron Image. Magnetic Focusing. Construction and Operation of Dissector Tube. Receiving Oscillite. Method of Synchronising. Comparisons. Review of Position in America.

Dr. V. Zworykin, formerly of the Westinghouse Electric and Manufacturing Company, and now associated with the R.C.A. Victor Company, to whom reference has already been made in this book, has of recent years interested himself in television and endeavoured to achieve satisfactory results by means of the cathode ray method, to which reference has also been made previously. Little information is available concerning the results which Dr. Zworykin has achieved and the exact methods used, the latest available information being, in fact, dated November, 1929, when Zworykin gave a brief description of his work before the Institute of Radio Engineers in New York. It is on this information that the following description is based. Some additional, but less up-to-date information, including circuit diagrams, is contained in British Patent No. 255,057, to which the more seriously interested reader is referred.

In order to provide a transmission wherewith to work, Zworykin uses a modified form of talking moving picture film projector, the essential elements of which are shown in Fig. 114. The intermittent motion device is eliminated from this projector, so that the film moves downward at constant speed. The light source consists of an ordinary 6-volt motor car lamp, the light from which is focused by the condenser lens on to the diaphragm, which has a small orifice in the centre. The emerging beam of light falls on the vibrating concave mirror, M, via a deflecting prism which is inserted for convenience and economy of space in the projector. The mirror, M, focuses the beam of light into a fine pencil point where it pierces the moving film.

With the mirror vibrating at a frequency of 480 cycles about a vertical axis, the light-spot scans the film horizontally, whilst a vertical scanning movement is obtained through the downward motion of the film. The vibrating mirror consists of a small steel rod with a vane, which is mounted between the poles of an electromagnet. The poles are U-shaped, and each leg is provided with a coil. An oscillating current of the same

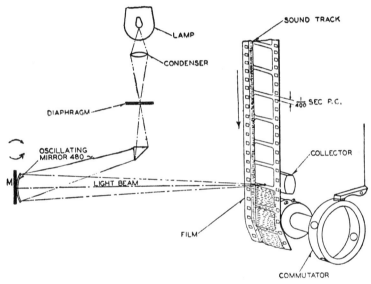

Fig. 114.—Essentials of Dr. V. Zworykin's talking motion picture film transmitter.

frequency as the natural frequency of the rod (480 cycles) is supplied to the coils, thus causing the rod and the mirror to oscillate about the axis of the rod.

Behind the moving film a lens, marked "collector" in Fig. 114, focuses the light beam, as it emerges from the film, on to a photoelectric cell. By this arrangement the scanning beam is always focused on a stationary spot in the cell, and adverse effects due to a lack of uniformity of the sensitive cathode surface are avoided.

A few moment's close study of this method of scanning will

reveal that the velocity of the beam across the film is not uniform. In the centre of the film the velocity is, in fact, about 57 per cent. higher than that of a spot (say, from a disc) scanning at uniform rate a picture of the same width. It was anticipated that this feature would prove objectionable, and it was planned to correct it by means of an optical filter. Actual tests revealed, however, that the non-uniformity of light distribution is not apparent to the eye.

The receiver used to operate in conjunction with this transmitter consists of a cathode ray tube specially designed for the purpose. The ordinary form of cathode ray tube, as used in oscillographs, is provided with means to cause the ray to scan in two dimensions, but no provision is made for varying the intensity of the ray. Furthermore, neither of the principal types of tube is suitable for television. The high voltage type, which would give sufficient brilliance for television purposes, is always operated in connection with a vacuum pump, which is obviously impractical for home use. The low voltage type, although sealed off, is altogether too lacking in brilliancy for television purposes. In order to give sufficient brilliancy for a picture measuring 5 inches square, the tube would have to be operated at a potential of at least 3000 volts. For larger pictures a still higher voltage would be required, because the brilliancy increases with the accelerating voltage.

Dr. Zworykin therefore developed a new type of tube, shown in Fig. 115. An oxide-coated filament is mounted within a controlling electrode, C. The stream of electrons shot off from the filament, which constitutes the cathode ray or beam, passes through a small hole in the front part of the controlling element, C, and then again through a small hole in the first anode, A. The first anode accelerates the electrons to a velocity of 300 to 400 volts. There is also a second anode consisting of a metallic coating on the inside of the glass bulb. This second anode accelerates the electrons still further up to 3000 or 4000 volts. At this voltage the velocity of electrons is about one-tenth that of light.

Another important function of this second electrode is to focus the beam electrostatically into a sharp point on the fluorescent screen. This screen is about 7 inches in diameter, and is covered with a fluorescent material such as willemite prepared by a special process so as to make it slightly conductive. The object of this slight conductivity is to permit the electrical charges (accumulations of electrons) set up on the screen by the cathode ray beam to leak off.

As has already been described, the stream of electrons which constitutes a cathode ray can easily be bent or moved in any direction by either an electrostatic or an electromagnetic field, and in the present case both methods are used, a set of deflecting plates and a set of deflecting coils being mounted in the same plane on the neck of the tube, on the outside of the glass wall. By setting these two deflectors in the same plane, the one electrostatic and the other electromagnetic, movement

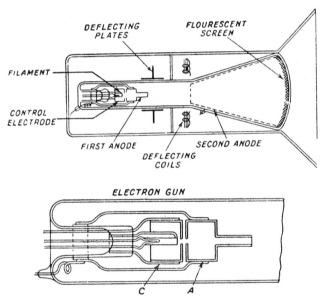

FIG. 115.—Cross-sectional drawings of Dr. Zworykin's cathode ray tube, or kinescope, for television.

of the beam in two directions at right angles to one another can be achieved.

The advantage of mounting the deflecting elements on the neck of the tube, between the first and second anodes, is that the deflecting field is caused to act on comparatively slowly moving electrons. Hence the field strength required is much less than that which would otherwise be required to deflect the beam under the full acceleration of the potential on the second anode.

The degree of brilliance of the line traced on the fluorescent screen by the moving cathode ray spot can be controlled to any desired extent by applying a suitable negative bias to the controlling element, C. The value of this bias then controls the mean intensity of the picture whose lights and shadows are then superimposed on this mean intensity by applying to the biassed control electrode the amplified picture impulses from the transmitter. In order to build up the picture, of course, it is necessary to deflect the beam in synchronism with the motion of the light beam across the moving picture film.

If separate channels are available for each of the synchronising signals the problem of synchronising the receiver with the transmitter is simple. To carry out synchronously the horizontal component of the scanning at the receiver it is only necessary to transmit to the receiver the sinusoidal 480 cycle current which is used to oscillate the mirror at the transmitter, and impress it on the deflecting coils of the cathode ray tube, or kinescope as Zworykin calls it. The cathode ray beam will then follow exactly the horizontal scanning movement of the light-spot across the film.

For vertical scanning, a voltage is generated at the receiver which is controlled by the incoming picture signals from the transmitter. To achieve this a condenser is charged at constant current through a current-limiting device, such as a two-electrode valve, so that the voltage at the condenser rises linearly. The deflecting plates of the kinescope are connected in parallel to this condenser, so that when the condenser is charging the cathode ray beam is deflected gradually from the bottom to the top of the fluorescent screen at constant speed. This speed is regulated by the temperature of the filament of the two-electrode charging valve to duplicate the downward movement of the film. An impulse is sent from the transmitter between pictures which discharges the condenser, thus quickly returning the beam to the bottom position, ready to start upward and reproduce the next picture.

For the transmission of the complete picture, therefore, three sets of signals are required: picture signals, horizontal scanning frequency, and impulses for framing (i.e. synchronising between individual pictures). Zworykin found that it is possible to combine all three sets of signals and send them over a single channel. To do this, the output of the photoelectric cells is first amplified to a sufficiently high level for transmission. There is then superimposed upon this signal a series of high

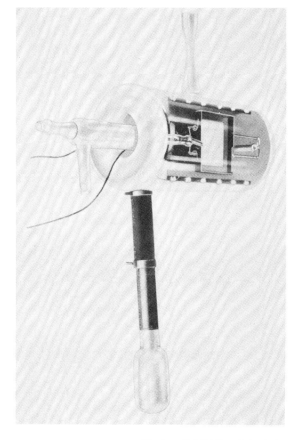

[*Courtesy: Bell Telephone Laboratories.*

The water-cooled neon tube used in the Bell system two-way television system. The glow takes place over the central rectangular portion. The hydrogen supply is in the lower extension tube.

[*To face page* 210.

PRESENT STATE IN AMERICA 211

audio frequency impulses lasting a few cycles only and occurring when the light beam passes the interval between the pictures. This high audio frequency is supplied by a 4000 cycle oscillator, connected as shown in Fig. 116. The circuit coupling this frequency to the image circuit is closed only during the intervals between pictures by the commutator shown connected to the film sprocket shaft in Fig. 115.

The picture frequencies, together with the superimposed framing frequency, are then passed through a band eliminating filter which removes the component of the picture frequency which has the same frequency as the horizontal scanning impulses, i.e. 480 cycles. Following this, a portion of the voltage which drives the vibrating mirror of the transmitter is impressed upon the signals, passed through the filter, and the entire

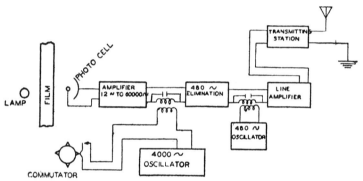

FIG. 116.—Circuit arrangements at the transmitting end.

spectrum of signals, extending from 12 to 60,000 cycles, is then caused to modulate the radio carrier wave, as indicated in Figs. 116 and 117.

At the receiving end of the circuit, the output of the radio receiver, after amplification, is divided by a band-pass filter into two parts, one the 480 cycle horizontal scanning frequency and the other the picture frequency plus the 4000 cycle framing frequency. The 480 cycle component is amplified by a tuned amplifier and then applied to the deflecting coils of the kinescope, as indicated in Fig. 118.

The picture and framing frequencies are applied directly to the control electrode (C in Fig. 115) of the kinescope. Part of this same energy is also impressed upon a band-pass filter which is tuned to the frequency of the A.C. voltage used for the

212　FIRST PRINCIPLES OF TELEVISION

framing impulses, i.e. 4000 cycles. The output of this filter is amplified, rectified, and used to unbias a discharging 3-electrode valve which is normally biassed to zero plate current, and which takes its plate voltage from the condenser which provides the vertical scanning voltage.

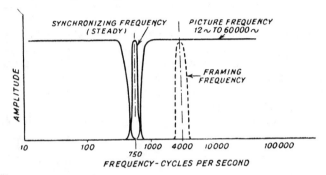

Fig. 117.—The frequency spectrum which is impressed on the radio carrier wave.

In this way the image signals and synchronising impulses are all transmitted over one channel and, according to Zworykin, fully automatic synchronisation is obtained.

The cathode ray tube method of achieving television offers a number of distinct advantages over all other known receiving

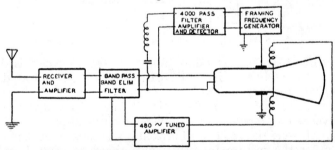

Fig. 118.—Circuit arrangements at the receiving end.

devices. One very valuable feature is the persistence of fluorescence of the screen, which, added to persistence of vision, makes it possible to reduce the number of individual pictures transmitted per second without causing noticeable flickering of the image. The standard rate of transmission used by Baird is $12\frac{1}{2}$ pictures per second, and is so low that flickering is very noticeable,

PRESENT STATE IN AMERICA

because persistence of vision only is relied upon to merge the individual pictures into a smoothly flowing moving scene. Using a cathode ray tube for reception, this speed of transmission is said to be entirely adequate. Within a fixed maximum frequency band width, therefore, a decrease in the number of pictures transmitted per second means that the number of lines into which individual pictures are divided can be increased, thus increasing the detail of the image.

By comparison with the conventional scanning-disc-neon-lamp receiver, a cathode ray receiver presents a number of differences. The image, as viewed on the end of the tube, is green instead of red, and it is visible to a larger number of people at once, for an enlargement of the image by means of lenses is unnecessary. There are no moving parts, and therefore no noise. The framing of the picture is automatic, and it is brilliant enough to be seen in a moderately lighted room. High power amplification of the incoming picture signal is unnecessary, for the power required to operate the control element of the kinescope is no more than that required to control the grid of an ordinary 3-electrode valve. Against these advantages, however, must be placed the necessity for providing the high accelerating voltages required for the anodes of the cathode ray tube. Where A.C. supply is available, however, this presents no technical difficulty, only a rather heavy item of expense. Then again, as has already been explained, cathode ray tubes are expensive as to first cost, and short-lived. However, it is highly probable that these disadvantages will be overcome in time.

Another American worker who has been extremely but unobtrusively active during the past few years is Philo T. Farnsworth, of Television Laboratories, Inc., San Francisco. Like Dr. Zworykin, Mr. Farnsworth is using the cathode ray tube for reception purposes. In addition, he is the only important worker known to the author who is making serious and successful efforts to make use of the cathode ray tube for transmission as well. Only the scantiest of information concerning the Farnsworth system has so far been published, and for the following detailed information the author is indebted to the courtesy of the inventor himself.

From a study of the structure of the human eye and what it is able to see, Farnsworth set about his television investigations on the premise that 200,000 picture elements are sufficient to give sufficiently pleasing detail to a picture, measuring $2\frac{3}{4}$ inches by 4 inches, giving a three-quarter length view of three people. Under these circumstances, if the method of scanning is as shown in Fig. 119, and 12 pictures are to be transmitted per second, then there will be two scanning frequencies, one a saw-tooth wave having a frequency of 12 cycles per second, and the other a similar wave having a frequency of 4800 cycles per second. The highest fundamental picture frequency will be 1200 Kc.

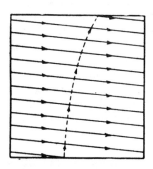

FIG. 119.—Farnsworth's method of scanning.

The form of cathode ray tube used by Farnsworth for transmission purposes is called by him an image dissector tube. It is shown diagrammatically in Fig. 120, and in the photograph facing page 228. Fig. 121 gives a cross-sectional view of the tube, which comprises a cathode, C, coated with photosensitive material

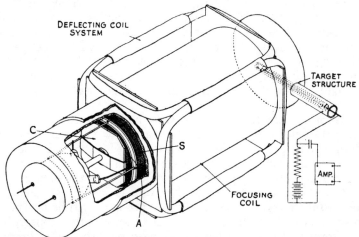

FIG. 120.—Details of the cathode ray or dissector tube used by Farnsworth for transmission.

and mounted parallel and close to an electrostatic shield, S. At the opposite end of the tube there is a target electrode having all but a single small area shielded from the discharge.

PRESENT STATE IN AMERICA

Considered broadly, this tube is a form of photoelectric cell wherein provision is made for forming an "electron image" of an optical image focused on the cathode surface. By "electron image," Farnsworth means that if a fluorescent screen were placed in the plane of the electron image, the original optical image would be reproduced. For this to happen, it is essential that every electron emitted from any single point on the cathode surface shall impinge on a corresponding point in the plane of the electron image.

The difficulties in the way of securing a sharply focused electron image may be understood with the aid of Fig. 122, where it is assumed that the image of a point of light falls on the cathode at the point P. If all the electrons leaving P could be constrained to travel parallel to one another, then any transverse section of the beam would represent a perfect image. But in order to achieve this desirable state of affairs it is essential that there must be no component of transverse velocity in the electron stream. These photoelectrons are emitted with a random velocity of from zero to about three volts. Another component of transverse velocity is added by the bending of the electrostatic lines near the wires of the anode screen, A, and still another component is added by microscopic irregularities of the cathode surface. The sum of these effects has been found to give a

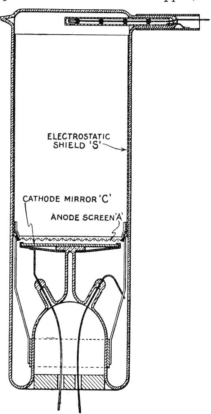

FIG. 121.—Cross-section of Farnsworth's dissector tube.

divergent pencil or cone of electrons from each point, the angle of which is about 5 degrees in the dissector in use at the time of writing.

For this reason, only a very poor electron image will result if an attempt is made to produce parallel electron paths. However, since the paths of the electrons are not distributed uniformly over the cone, it is possible to obtain something of an electron image in the plane, T, by the use of low-frequency light, careful construction of the anode screen, and the use of high potential between the cathode and anode. The dissector tube was, in fact, first used in this manner, but it has been found possible to focus the electrons magnetically to give an electron

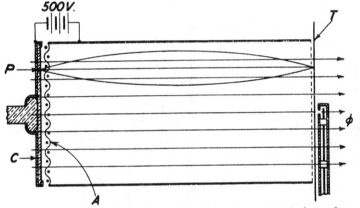

FIG. 122.—Illustrating the difficulty of securing a sharply focused electron image.

image which is fully comparable in sharpness to an optical image.

Magnetic focusing is accomplished by applying a magnetic field of the proper density in such a manner that the lines of force are parallel to the axis of the tube. The effect of this magnetic field is to bend the electrons into helical paths tangent to the line of magnetic force through the emitting point. Hence, since all the electrons start on a line of magnetic force they will arrive back at the same line of force when they have traversed the circle once.

The axis of the helices need not be within the electron pencil, for if the longitudinal velocity is the same for all, every electron will eventually return to tangency with a line of magnetic

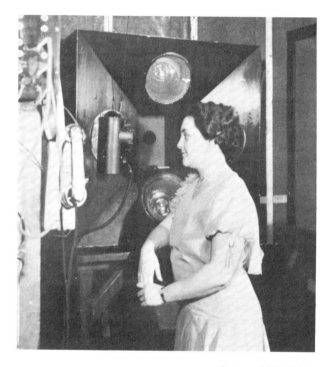

[*Courtesy: G.E.C. (U.S.A.).*

The experimental television transmitter used by the G.E.C. during the demonstrations of Dr. Alexanderson's large screen. Four large photoelectric cells were used, one on each of the four sides of the opening leading to the spotlight. The artiste is facing the microphone.

[*To face page* 216.

PRESENT STATE IN AMERICA

force passing through its point of origin. Thus, if the direction of the field is changed, the point at which the rays are focused will follow the field, so that if a transverse magnetic field is superimposed on the longitudinal field, the electrons will be deflected to follow the resultant field, as shown in Fig. 123. Hence, if a pair of coils is placed one on each side of the dissector tube, and energised with an alternating current, they will give to the electron stream a deflection along the axis of the coils, or in a direction parallel to the magnetic lines produced by them, rather than at right angles to this direction, as would be the case if there were no longitudinal focusing field.

In this manner Farnsworth has overcome one of the greatest

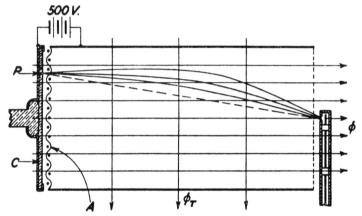

FIG. 123.—Illustrating how the electron streams are deflected on to the aperture leading to the "target" or anode of the photoelectric cell.

difficulties encountered by those who have attempted to utilise cathode ray tubes for television purposes, but, having achieved such a degree of perfection with carefully made tubes that a stationary electron image is not inferior to a good optical image, there remains the slight blurring effect at the edges which occurs when this image is moved by the scanning operations. The reasons for this blurring in the present case are: (1) The distance between the aperture in Fig. 123 and the cathode is greater for the edges than for the centre of the image. (2) The electron velocity in the direction of the aperture is less for the edges than for the centre. (3) The resultant magnetic field in the direction of the path of the electrons increases as the angle from the centre increases. All these factors can be reduced by increasing the deflection distance, and in practice 15 degrees deflection on either side of zero is the value used when the scanning aperture is not smaller than 0·015 inch.

The high vacuum type dissector tube illustrated in Figs. 120 and 121 consists of a cylindrical glass tube, having at one end a flat window which is polished before sealing in. At the other end is a stem upon which the elements of the tube are supported, and through which the leads pass. The inner end of the stem carries a short glass pillar which terminates in a square button. This button supports a silvered mirror upon which is deposited a photosensitive film. A band clamp supported by the stem carries the anode structure. This anode structure is made by winding a very fine tungsten wire around a thin flat tungsten-nickel frame, and it is supported so that it is closely parallel to the cathode. In the latest types of tubes the electrostatic shield, S, Fig. 121, consists of a platinum coating on the inner walls of the tube. The target is supported from a side tubulation at the end opposite the cathode. This target consists essentially of a small nickel anode tube which acts as a target for the arriving electron, or cathode ray stream, which is so focused that it reaches the anode tube through a tiny aperture in the encasing wall of the glass tube.

In operation, an image of the object to be transmitted is focused through the polished window on to the cathode. Electrons are emitted from the photosensitive cathode mirror in proportion to the amount of light falling on it, and these electrons are accelerated by a potential of the order of 500 volts which is applied between the cathode and the anode screen, as shown in Fig. 122. Most of the electrons are projected into the equipotential region between the anode screen and the target, wherein they follow a helical path as already described, and recombine to form an electron image in the plane of the target. This electron image is then shifted by the transverse magnetic field, so that the entire image is caused to move over the aperture in the target shield, thus achieving the scanning of the image.

This transverse scanning field is produced in actual practice by two sets of coils which are mounted at right angles to one another on the outside of the dissector tube, and outside the focusing coil winding, as shown in Fig. 120. A saw-tooth wave alternating current of about 3000 cycles flows through one set of coils, and produces a horizontal deflection of the image. A 15 cycle current of similar wave form flows through the other set of coils and produces a vertical deflection of the image. The resultant path of the image, relative to the target aperture, is as shown in Fig. 119. Thus, each individual picture is scanned in 200 lines, and the scanning time for one line is 1/3000th of a second. The amplifier which handles the output of the target element of the tube must therefore be capable of handling a band width of approximately 300 Kc.

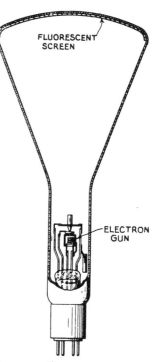

FIG. 124.—Cross-section of Farnsworth's receiving cathode ray tube, or oscillite.

The design and construction of an amplifier to handle such an enormous frequency band width is a gigantic problem in itself, but Farnsworth has solved it by making use of what he terms a system of admittance neutralisation, which permits input impedances (as well as interstage valve impedance) of as high as several megohms to be obtained up to one million cycles or more. At the time of writing he is using an amplifier with a frequency characteristic which is approximately flat up to 600 Kc. Unfortunately, the patent situation does not permit of a description of this amplifier, and its underlying principle, being published at this time.

At the receiving end the incoming picture impulses are transformed into a visible image by means of a form of cathode ray tube called by Farnsworth an oscillite. This is illustrated diagrammatically in Fig. 124, details of the electrode structure being given in Figs. 125 and 126, and a photograph of the tube appears facing page 228.

The oscillite is similar in some respects to Zworykin's kinescope, but makes use of the magnetic focusing principle, and scanning is effected by means of two sets of coils at right angles to one another, as in the transmitting dissector tube.

220 FIRST PRINCIPLES OF TELEVISION

The electron gun element has been designed with the object of driving the greatest possible number of electrons through an aperture of given size, and limiting the angle of this beam so that it can be easily focused. This element, shown in Fig. 125, consists of a helical filament oxide-coated only on the inside. A shield is placed over this filament having in it a hole the same diameter as the filament helix. The anode, which is tubular in form, is positioned in front of the cathode, and midway

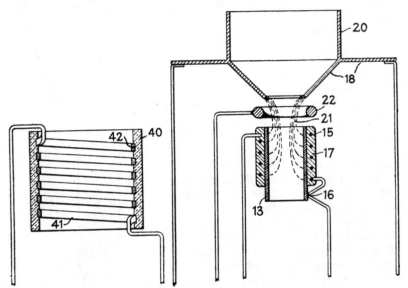

Fig. 125.—Details of the filament of the oscillite.

Fig. 126.—The electron gun, showing filament (15), electron stream (21), control element (22), and tubular anode (20).

between the filament shield and the anode there is mounted a ring grid, marked 22 in Fig. 126.

The advantage of this type of element is that the anode tube is mounted approximately at the focal point of the electrons leaving the emitter, or filament. The anode potential required to provide this focal point at the entrance to the anode tube may be any voltage between 1500 and 2500 volts for the tubes in use at the time of writing.

An interesting point which has been observed when using these tubes is that they function only when secondary electrons are emitted from the fluorescent screen. Sometimes a black

PRESENT STATE IN AMERICA

spot will appear on the end of the tube, caused by that point charging up negatively. This effect is due to the fact that unstable conditions exist at this point, which may assume a large positive or negative charge with respect to the anode. According to Farnsworth, the effect is not bothersome at all, for it is necessary to have a very high current density in order to observe it.

In order to achieve synchronism, Farnsworth generates at the receiver two alternating currents of saw-tooth wave form (Fig. 127) identical to those used at the transmitter. These currents, of course, have to be synchronised with those at the transmitter. In order to do this, advantage is taken of the fact that the currents can be made to induce a strong voltage pulse into the picture frequency circuit during the steep part of their slope. These pulses are used at the receiving end to

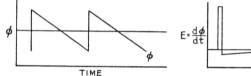

FIG. 127.—Wave form of deflecting coil current.

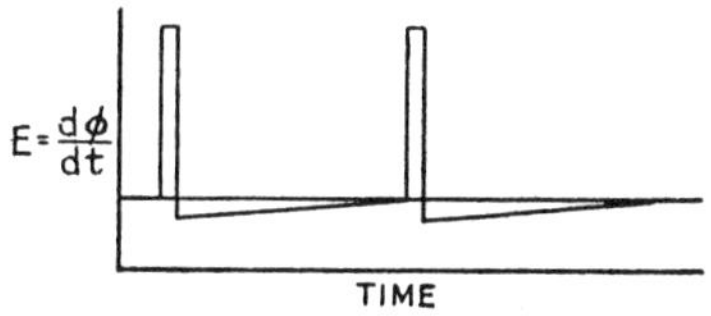

FIG. 128.—Voltage impulses applied to output of amplifier and to line between individual pictures, and proportional to the rate of change shown in Fig. 127.

hold the local generators in step. These voltage pulses, which occur during the interval between individual pictures (see Fig. 128), serve also to turn off the oscillite spot during the return part of its path, i.e. during the very steep part of the saw-tooth wave cycle.

This system of synchronisation, according to Farnsworth, is very simple and very effective. It does not require an additional communication channel for the synchronising impulses, nor even any additional equipment such as filters, etc., to separate the synchronising impulses from the picture frequency.

The reason for using a saw-tooth wave form in preference to a pure sine wave for energising the deflector coils is that if a sine wave current were used, a double picture would be produced at the receiver whenever the two currents were not in phase. The reason for this may be seen from Fig. 129, which represents a picture of a single line. It will be seen that when the two currents are slightly out of phase the scanning point is shifted towards one side when it is travelling across the picture in one direction, and towards the other side for its return trip. This effect is completely eliminated with saw-tooth wave scanning.

The scanning frequencies are generated by means of a glow tube in combination with a small power valve used as an oscillator, and one stage of amplification. The glow tube has an electrode sealed into it which is coupled through a 10 mmfd. condenser to the picture frequency circuit. It is found in practice that the pulses present in this circuit lock the receiver oscillator tightly in step with that at the transmitter. A diagram of the receiver arrangements is given in Fig. 130, which

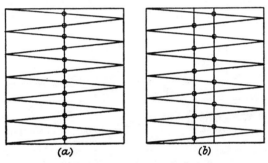

FIG. 129.—Illustrating why saw-tooth instead of sine-wave currents are used for synchronising. (a) Two synchronising currents in phase; image clear. (b) Using sine-wave form; currents slightly out of phase; double image produced.

shows the wireless receiver (41) and the connection to the electron stream control element (47), and one of the two oscillator circuits which supply the deflector coils (30). The power required in the deflector coils for maximum size pictures may be generated by a small power valve. The amount of power required for the focusing coil is quite negligible.

Considerable work has been done by Farnsworth on the development of a 4-metre wireless link, and the progress to date indicates that a quite satisfactory television service could be conducted on this wavelength over distances up to about 50 miles, providing proper care is taken in choosing the location of the transmitter. This involves so placing the transmitter that it will be almost visible from any part of the area it is

[*Photos Courtesy: G.E.C. (U.S.A.)*

Above: General view of Dr. Alexanderson's large screen apparatus. Note loud speakers on each side. When in use curtains are dropped about the sides and top of the screen, which measures 6 by 7 feet. *Below:* Dr. Alexanderson at the synchronising control of his large screen projector. See Figs. 112 and 113.

[*To face page* 222.

PRESENT STATE IN AMERICA

designed to serve, for only the ground or direct component of the wave can be relied upon to affect the receiver.

More successful results have been achieved by using wired wireless as the channel of communication. It has been found to be perfectly feasible to modulate a 1000 Kc. carrier wave with a 300 Kc. band width, and transmit it over a telephone line. The pictures so transmitted are practically as good as those seen on a monitoring receiver placed close to the transmitter. The voltage attenuation has been found to be about

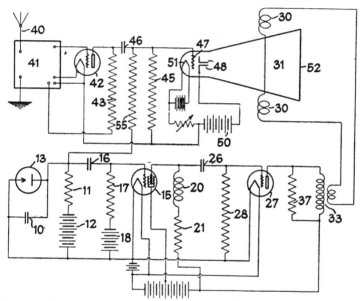

Fig. 130.—Circuit diagram showing wireless receiver (41) and output valve (42) which connects with the control element (47) of the oscillite. The lower part of the diagram shows the circuit of one of the synchronising current generators.

45 db. per mile for a standard American No. 19 pair cable. With this attenuation it would probably be necessary to relay every few miles, but if an open wire line were to be used instead, Farnsworth is of the opinion that a television service by means of wired wireless would come entirely within the range of possibility. The synchronising impulses are, of course, also sent along the same telephone line, and voice frequencies may be put on the same line as well, thus confining the entire " see and hear " service to a single channel of communication.

A frequency band 300 Kc. wide cannot, of course, be broadcast through the medium of an ordinary sound broadcasting station. Quite apart from the technical difficulties involved, broadcasting stations are confined to a channel only 10 Kc. wide in the United States, and 9 Kc. wide in Europe.

However, having thus very ingeniously developed a special amplifier for very high frequencies, and means for transmitting those frequencies both by radio and by wire, Farnsworth has recently announced the invention of what he calls an "image compressor." Details of this have not yet been released, but it is understood that by means of this new development, part of the wide picture frequency band is suppressed before transmission so that the width of the frequency band which must actually be transmitted is only about 7 Kc. The suppressed part of the signal is replaced locally at the receiver. This development, like most of his other inventions, was worked out by Farnsworth mathematically first, and then put into practice. It is his brilliance as a mathematician that has enabled this young Californian to make such good progress.

By applying the new transmission principle Farnsworth hopes ultimately to be able to transmit television signals on one side band of an ordinary broadcasting station, and speech on the other, thus providing a complete seeing and hearing service through the regular broadcasting stations. Farnsworth states that his images, when received by the "image compressor" method, are even better than they are when transmitted in the ordinary way by wire or radio, using his special wide band amplifiers. The images measure about 4 inches square, unmagnified.

Farnsworth estimates that a complete unit, comprising television receiver and all-mains radio receiver will cost about £50. An adaptor, comprising oscillite, synchronising unit, and 2-valve L.F. amplifier, suitable for attachment to existing radio sets, should sell for about £20. The average life of the oscillite is 1000 hours, the cost of renewal £2 to £3, and the operating high voltage, obtainable from an H.T. eliminator, is 2000 volts. The essentials of the equipment are thus seen to be simple, reasonably cheap, and absolutely silent in operation.

Pursuing his experiments still further, Farnsworth has, at the time of writing, succeeded in increasing the brilliancy of the fluorescent screen of his oscillite to such a degree that the image is now too bright to be viewed directly on the end of the tube. It is, indeed, surprising that fluorescent materials have been made to glow so brightly and yet preserve a reasonable length of life. Under these circumstances Farnsworth has placed a condensing lens in front of the fluorescent screen, and now projects his image on to a ground glass screen measuring 2 feet square.

When a disc scanner is in use there is a theoretical advantage in increasing the number of holes in the disc (scanning lines), but in practice there is actually very little improvement to be observed in the image by so doing. A 60-line image looks very little better than a 30-line image. But when it comes to enlarging the image by projection on to a screen, an immediate coarsening of detail results from the magnification, and an increased number of scanning lines is a decided advantage. As a result of his projection experiments, therefore, Farnsworth has increased the number of his scanning lines from 200 to 400. At a transmission frequency of fifteen pictures per second this means a total picture frequency of 1200 Kc., which Farnsworth states he can transmit by radio, using his "image compressor" system, on a frequency band less than 15 Kc. wide. Standard L.F. transformers are used to handle the signal currents.

An advantageous feature of the Farnsworth system is that changes in the number of scanning lines, and in the number of pictures transmitted per second, can be made with the greatest of ease, by simply varying the frequency of the two A.C. currents supplied to the deflector coils. Thus, the Farnsworth television receiver may be termed a universal receiver, capable of picking up anybody's transmissions (so long as horizontal scanning and a square picture are being employed) irrespective of the number of scanning lines used and the number of pictures transmitted per second.

To come now to a comparison of results, the Bell Telephone Laboratories have produced by their mechanical methods the finest images that the author has seen to date. They are adequately brilliant, extremely steady, and contain an amount of detail comparable with a moving picture close-up of the artistic diffused-focus type. When talking over the two-way television circuit the illusion that one is talking to a man just across the room is wellnigh perfect. This illusion is aided considerably by the perfection and skilful arrangement of the acoustic circuit. The image is not distorted by over-magnification, but just sufficiently enlarged to appear to be about 5 or 6 inches square. Even at the rate of transmission of 18 pictures per second, however, there is still a noticeable amount of flicker, which is comparable with the flicker of the early motion picture. Another noticeable feature is that although the present 72-hole discs give double the number of picture elements produced by 50-hole discs, the increase in the detail of the image only amounts to approximately 50 per cent. This is largely due to causes discussed in Chapter X.

Alexanderson's large screen is comparable in detail with Karolus' 3-foot screen, but is very much more brilliant as regards illumination. By comparison with Baird's large 2100 element screen, it is not so brilliant, but the detail is better. As regards swaying propensities, due to synchronising errors, they are both about the same.

It has not been possible for the author to see Zworykin's image, as no demonstrations are being given at the present time. Neither has he been able to see Farnsworth's images, since to witness a demonstration would entail a trip to San Francisco, but judging by the accounts of responsible witnesses who have seen demonstrations, and from an examination of photographs of Farnsworth's images (which, unfortunately, are so poor photographically that they cannot be reproduced here), the author concludes that the Farnsworth images compare very favourably with the Bell images. There is, according to witnesses, no visible movement due to synchronising errors.

The Jenkins' 48-line images are not so good as Baird's 30-line images on his small receiver, and the 60-line images show but little improvement. This is explained partly by the fact that the commercial Jenkins televisor is equipped with too powerful a magnifying system in an attempt to make the image as large as possible. This magnifying lens, besides showing up defects in the image, distorts the sides of it. Also, as already explained, horizontal scanning does not give such good results with a head and shoulders image as does vertical scanning. Baird's picture ratio of 7 to 3 favours a better image than the 1 to 1 ratio employed by Jenkins, for it makes possible the achievement of greater detail in the lengthwise scanning.

The present position of television in America may be summed up as follows :

The Jenkins Television Corporation, believing that a start

must be made somewhere in the commercialisation of television, and that the means at present available justify a start, are, as already stated, broadcasting experimental programmes on regular schedules, and endeavouring to interest the public sufficiently to purchase television receivers or kits of parts.

The Bell Telephone Laboratories, maintained by the American Telephone and Telegraph Company, maintain their traditional attitude, which is that they are interested in any new development of electrical communication, and are willing to develop it to the ultimate limit in the hope that eventually they will be able to apply it to their particular sphere of interest, which is universal commercial communication over wires or by wireless. They are not directly interested in the broadcasting of television any more than they are directly interested in the broadcasting of sound. Any system of television which they may develop, therefore, will not be readily available to members of the public for use in their own homes, except, perhaps, at some future date, as an adjunct to the telephone to enable the distant correspondent to be seen. The equipment in use at present is obviously too complicated and expensive for home use.

The Radio Corporation of America has recently entered the entertainment field by associating itself with the Keith-Orpheum cinema and vaudeville interests, the combination being known as R.K.O.—Radio-Keith-Orpheum. In addition, the R.C.A. has purchased the Victor Talking Machine Company (His Master's Voice), this combination being known as the R.C.A. Victor Company. Since the General Electric and Westinghouse companies have interests in the R.C.A., it was with no surprise that the author, on starting out to investigate the American position, learned that the entire television staffs of the General Electric and Westinghouse Companies, including Dr. Zworykin, but not Dr. Alexanderson, have been merged and transferred to the R.C.A. Victor plant at Camden, N.J. Although the author enjoyed a very interesting discussion with Dr. Zworykin and members of his staff, it was not possible for them to disclose anything concerning the work that is being done, as it is being carried on in secret for the time being.

The new Chicago studios of the National Broadcasting Company, which is a subsidiary of the R.C.A., have been so designed that they can very readily be adapted to the needs of television when it comes along in a sufficiently perfect form to be incorporated in the regular broadcast programmes as an added entertainment feature. In New York, work has been started on a £50,000,000 building project, to be known as Radio City, which when completed will be the New York headquarters of the National Broadcasting Company. Mr. M. H. Aylesworth, President of the N.B.C., has informed the author that the studios in this Radio City will also be arranged to accommodate television. It is, therefore, not difficult to deduce that the R.C.A. expects television to be in a form acceptable to the public by the time Radio City is ready (which will be about 1934 according to present estimates), and, further, that the R.C.A. intends to be the prime mover in the development of television to that point, and to control it commercially afterwards. Both the N.B.C. and its rival, the Columbia Broadcasting System, have experimental television broadcasting licences (see Appendix to Chapter XIV.).

Meanwhile, Mr. Farnsworth continues to forge ahead rapidly and scientifically, and bids fair to become a serious rival to the R.C.A. interests unless he chooses at some later date to associate himself with them, whereupon, as the situation appears at present, the R.C.A. will have a virtual monopoly of television as applied to entertainment, just as to-day they have a virtual monopoly, through their patent holdings, in the radio field.

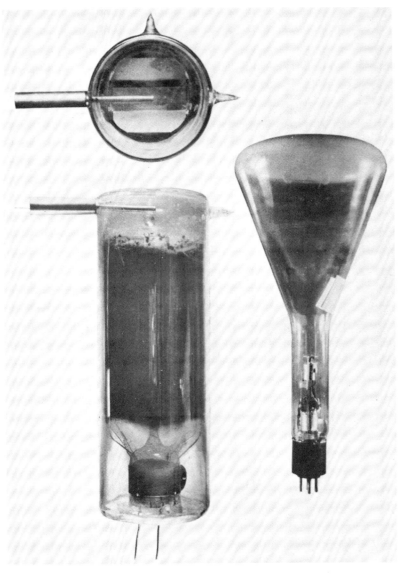

[*Courtesy: P. T. Farnsworth.*

Farnsworth's image dissector (left), with end view of tube at top. Right, the oscillite.

CHAPTER XVI.

CONCLUSIONS.

Various Opinions Concerning Television. Possibilities of Television as an Entertainment Medium, and Requirements. Can it be Done? Unlimited Scope for the Amateur. Does the Public Want Home Television? Television as a Stepping-Stone, and as an Adjunct to the Motion Picture Industry. Other Applications. Noctovision and Television as Aids to Navigation. The International Newspaper.

During the last few years those interested in the exploitation of television have been responsible, both in Europe and America, for the release in newspapers and magazines of a flood of publicity designed to convince the public that " television is here," or " just around the corner." As one result, potential purchasers of wireless sets have postponed their purchases until such time as they can buy the long-promised set which will also incorporate television. This has caused grave concern to the Radio Manufacturers' Associations of both Great Britain and America. Both associations have been moved to discourage television demonstrations and exhibits at their annual radio exhibitions, and the American Association has even gone so far as to issue counter publicity denying that television is anywhere near ready for public consumption.

In the issue of the *New York Sun* for August 2nd, 1930, a writer by the name of R. P. Clarkson, commenting on this counter publicity, says that time has justified this action. " We have no television in the home and we are nowhere near having any. Recently the same old plugging away with the loose and misleading use of the term television at a time when radio is at a low tide has been noticeable. Now comes the backwash. The general engineer of the R.C.A." (Dr. Alfred N. Goldsmith) " makes the statement that television within a year would be a miracle. Of course, he is thinking solely of the experiments being made by his own staff. Within two years would be considered an amazing feat. Within five years might be considered possible. I'll go one better. If television is ever developed along present lines, I'll eat three scanning discs and a couple of kino " (neon) " lamps."

This refreshingly forceful and typically American method of expressing the situation represents the opinion of a very large body of qualified observers. C. W. Horn, general engineer of the National Broadcasting Company, describes television in its present form as being as far removed from perfection and entertainment value as was that forerunner of the modern motion picture, the old book of photographs, each slightly different from the other which, when thumbed over quickly, gave a crude motion picture effect.

Mr. Frank B. Jewett, President of the Bell Telephone Laboratories, in commenting upon that institution's two-way television system, says : " Although, on account of its present complexity and high cost, no substantial commercial field is yet in sight for television requiring good images, there is still a large amount of technical work which gives promise of decided improvements over the means and methods now available. Both because of this fact and because of the collateral influence which research and development work in the television field has on our general communication problems, Bell Telephone Laboratories will continue to explore the field of television."

In contrast to this conservative expression of opinion, Dr. Lee DeForest, the inventor of the audion or 3-electrode valve, expresses the following opinion in the final chapter of a book entitled " Radio and its Future," by Martin Codel. He foresees television developing along lines similar to those taken by radio, namely, "first, the experimental days, with the experimenters taking part with home-made equipment ; then the gradual crystallising of a practical system based on knowledge gained in actual work ; followed by the mobilisation of the essential capital, personnel and production facilities for the creation of an industry as well as the founding of a national institution."

As to the possibility of television competing with the motion picture, Dr. DeForest's answer is : " I cannot conceive of television eliminating the motion picture. The two serve totally different ends. Television . . . will be the spontaneous presentation—flashed before the audience for immediate enjoyment and enlightenment—born and dead in a fraction of a second. The motion picture . . . is recorded permanently and made available to any audience at any time and any place.

CONCLUSIONS

Just as the gramophone industry has joined hands with the radio industry, so must television and the motion picture join hands in the future."

It is perhaps desirable at this point to consider the possibilities of television as a means of entertainment. Let us set down what a television entertainment service will probably have to be capable of doing, and then see whether we are ever likely to be able to do it.

In the first place, television in its present form, showing only a head and shoulders view of a single individual, and that imperfectly, has definitely no entertainment value to persons other than those who have some direct or indirect scientific interest in the subject. Once the novelty has worn off there is nothing in the present limited images which will interest, amuse, or instruct the average non-technical person.

Television has come along at a time when high standards of performance have already been set up by the talking motion picture, both as a public and as a private means of entertainment. Consciously or unconsciously the public will expect, and probably demand, that television do all that the motion picture can do, and as well. That means that the equipment of the future, either large screen for public entertainment, or small screen for home use, will have to be capable of embracing a field of view, either indoors or outdoors, as extensive, complete, and flexible as is at present coverable by means of a motion picture camera, and the detail presented will have to be comparable with that of the motion picture also.

The home screen of the future, upon which the television images are to be depicted, will have to be, for comfort and convenience, not smaller than 9 by 12 inches, unmagnified. Magnifying lenses, besides distorting the picture and causing loss of light, limit the number of observers to two or three, and force even that number to place themselves, often in discomfort, directly in front of the lens. The standard of comfort of the home cinema must prevail on this point.

Having thus set the standards of field of view and size of screen, we can speculate a little further and attempt to fix quantitatively the amount of detail which must be provided. If scanning is to continue to be carried out in parallel strips or lines, it is probable that we shall not be satisfied with less than 100 lines per linear inch. That means that, for a 9 by 12 inch receiving screen, unmagnified, we shall have to resolve the image into 900 × 1200 = 1,080,000 picture elements. If we set the transmission speed at 15 images per second—a rather low figure—the frequency of the picture impulses which we must transmit, either by wire or by wireless, works out at 8100 Kc. We have already seen that it is quite impossible at the present time to transmit such an enormously high frequency, and a few minutes' thought, plus a little calculation, shows that we should need a scanning disc nearly 300 feet in diameter to fulfil the requirements set out above! The whole conception is preposterous and ridiculous.

This example may serve the purpose, however, of revealing the deadlock which television has now reached. One is tempted to say that "perfect" television, as outlined above, will always be impossible. But in scientific matters to-day's impossibilities have an uncanny knack of becoming to-morrow's commonplace achievements. Just because no known system is capable of meeting these requirements to-day, it is not safe to say that no system will ever be able to meet them. Some purely mechanical system may yet be developed which will meet the requirements. Some development of the cathode ray system may yet be completely successful. Or some entirely new system, as yet unheard of, may accomplish the desired end in such a simple manner that, when it appears, we shall exclaim: "Why ever did not somebody think of that before!"

Just as amateurs made a large number of important contributions to the development of the radio art, so they may yet contribute to the development of television. Enough has been written here to indicate the vast field which is open to them.

There is still another unknown quantity to be speculated upon. Does the public really want television as an adjunct to sound broadcasting? Does it really want this additional form of home entertainment? One may argue that the public did not want talking pictures when they were first made available, but they are now universally accepted.

But television as an adjunct to sound broadcasting is another matter. Consider the average home to-day, equipped with a modern all-mains wireless set. There is an increasing tendency to leave the set switched on for hours on end, without any member of the family paying any particular attention to it. The music is just "there," as a subconscious background. Every member of the family talks, reads, or engages in other pursuits. Occasionally, when something specially important or interesting is being broadcast, some or all of those present will listen. And even then, how often must there be an argument before somebody will stop talking! This state of affairs is more particularly general in America than it is in this country, but there are signs that we are following in America's footsteps in this as in most other radio broadcasting matters.

The reason for this state of affairs is not far to seek. Broadcasting requires the attention of but one of our senses, that of hearing, and we can hear a lot of things (imperfectly, no doubt, but often sufficiently well for our purpose) whilst occupied with other matters. Add to the loud speaker a vision screen, and what happens? We are called upon to use the sense of sight in conjunction with that of hearing, and to do so calls for the suspension of all other activities while we concentrate on the screen and the loud speaker. It is not difficult to imagine that the television part of the programme would receive but scant attention over long periods.

However, it is inadvisable and even dangerous to allow one's speculations on the future of a new idea to run riot. Broadcasting is just eleven years old in America. No one, eleven years ago, could foresee this spectacular development of the then embryonic art of radiotelephony. There is considerable food for thought in the following public statement made by Dr. Alexanderson :—

" Looking back over the development of the electrical industry we can clearly trace how the science of electricity gave birth to the electrical industry ; how later the electrical industry took hold of another branch of science and created the radio industry. We are able to some extent to project into the future the working of the natural social forces that give birth to new epochs, but as to the destiny and significance of these new movements after they have been launched, the engineer is peculiarly blind. Mr. Owen D. Young " (Chairman of the Board of Directors of the General Electric Co., and author of the famous Young Plan) " has repeatedly said that he has the great advantage of not being handicapped by scientific knowledge. His predictions of the future have also been much more far flung and correct than those of the engineers associated with him.

" For fifteen years radio was simply an auxiliary to navigation. In 1915 and 1916 we held daily communication by radio telephone from Schenectady to New York. We found that many amateurs adopted the habit of listening in, and our noon hour of radio became the first regular broadcasting. But we had no idea what it would lead to. Our idea was to telephone across the ocean, and so we did at the close of the war, but we failed to see the great social significance of broadcasting.

"Television is to-day in the same state as radio telephony was in 1915. We may derive some comfort from this experience of the past but, on the other hand, we are not sure that the analogy is justifiable and that television will repeat the history of radio telephony. We must then fall back upon our conviction that the development of television is inevitable on account of the forces working in the scientific world to-day, and that it is a satisfaction to make one's contribution to this evolution even if, in this case, the results should prove to be only a stepping-stone to something else."

There is considerable significance in that last phrase, a stepping-stone to something else. There are things other than attempts to provide home entertainment that television has already done, and unlimited possibilities for doing other things in the future. For example, the needs of television have accelerated considerably the development of the photoelectric cell, which has a thousand and one applications in other branches of industry. The needs of synchronism have caused to be developed electric motors which will run at practically constant speed, and control equipment which will cause them to run at absolutely constant speed. There are many uses to which constant speed motors can be put. Because of the needs of television we have learned a great deal more about gas discharge lamps, and we have developed amplifiers which will not only magnify extremely minute currents to a hitherto impossible degree, but will also handle, without distortion, hitherto impossible frequencies.

These are some of the things which television development has already produced, and which may be termed valuable by-products.

There is room for the further development of television itself as an adjunct to the motion picture industry. Many copies of films must now be made and distributed round the country. A sufficiently perfected system of television would enable one copy to be run off at a central distributing station, scanned and converted into electrical impulses, and transmitted either by wire or by wireless to subscribing theatres all over the country,

[*Courtesy: P. T. Farnsworth.*

An experimental television receiver built by Farnsworth. The image is seen on the circular end of the cathode ray tube, or oscillite.

[*To face page* 234.

CONCLUSIONS

which would transform the incoming signals into light and project it on to a screen. Important topical events, occurring at a convenient time, might be similarly transmitted by direct pick-up from the scene and shown to cinema theatre audiences all over the country.

The circulation of topical news films is to-day restricted by the time required for transportation. The direct pick-up of a topical event in London might be flashed across the Atlantic and, if the difference in time did not make such a course inconvenient, immediately presented to theatre audiences in New York. If the time difference interfered, the received signals could be recorded on film which could then be run off for exhibition at a more convenient time. That the motion picture industry is alive to the possibilities of television is evidenced by their great interest in it, and by the fact that producers, when buying scenarios or the film rights of books, now include in the contract the television rights.

Apart entirely from the entertainment side of television, the new science has many other applications, only a few of which will be mentioned here. For example, a determined beginning has yet to be made to develop the possibilities of noctovision as an aid to marine and aerial navigation during foggy weather. Various attempts, more or less abortive so far as general application is concerned, have been made from time to time to use some form of signalling through fog by means of an infrared ray beam, but so far, apart from some experiments which Baird and others have made, no determined effort has been made to produce a practicable system, capable of wide and general application on ships and aircraft, whereby the navigator can actually see through fog and thus avoid obstacles. But that will come.

The well-known American inventor, John Hays Hammond, Jr., has already described a projected system whereby, by a combination of television and direction finding, the pilot of an aeroplane approaching an airport in fog can see on a small screen on his dashboard, an image of the airport and its surroundings, together with his own position over it. This is done by televising, on the ground, a model of the airport on which the constantly changing position of the plane, as indicated by cross bearings taken of wireless signals emanating from the plane, is marked. The television signals are then wirelessed to the plane. This, of course, is a television method, pure and simple, of overcoming fog ; it does not make use of infra-red rays.

Dr. J. Robinson has patented a system (British Patent No. 327,112) which combines television and direction finding, so that the captain of a ship can himself read off visually his bearing from any given radio beacon station equipped for the purpose. Essentially, this method consists, at the transmitting station on shore, of a mechanically rotated frame aerial to which is coupled, also mechanically, a pointer travelling round a compass card. This is arranged so that when the pointer reads North and South, the frame aerial is orientated in a North and South direction. The compass card, with its revolving pointer, is then scanned by a television transmitter, the signals of which are then radiated by the frame aerial.

On board the ship, a simple receiver, working on an ordinary open aerial, picks up the signals and feeds them to a televisor. Naturally, as the transmitting frame aerial turns steadily round, the received signal strength will rise and fall, depending on whether or not the frame is pointing to the ship or at right angles to it. As a consequence, the image of the compass card and pointer, as seen on the receiving screen, will be clearest when the transmitting frame aerial is pointed at the ship, and invisible when it is at right angles to it. By noting the compass bearing, as indicated by the pointer, when the image is clearest, the captain learns his exact bearing from the transmitting station. Tuning in to a second station will give a second, or cross bearing, and the intersection of the two, when plotted on a chart, will give him his exact position.

Several automatic direction-finding systems, within which category Dr. Robinson's invention falls, have been invented from time to time, the chief objection to them being that they cannot differentiate between interfering stations, which might cause false readings. Dr. Robinson overcomes this objection by applying the Stenode Radiostat principle to the wireless receivers.

There is another potential field for television in connection with the world-wide disseminaton of visual news, or the simultaneous publication at very remote points of a newspaper, each remote edition being an exact duplicate in every particular of the original or central edition. To achieve this it would be necessary to send facsimile copies of each page, *en bloc*, to the remote points. Phototelegraphy can go part way towards the accomplishment of this, but it cannot yet deal with so large a surface as the opened page of a newspaper, nor can it deal with it in sufficient detail, nor sufficiently rapidly. A combination of the techniques of phototelegraphy and television would seem to be required, a kind of speeded up phototelegraphy, or a slowed down form of television, giving the amount of detail which is at present possible with the best systems of phototelegraphy.

An entire newspaper would have to be transmitted within, say, an hour, and recorded photographically, or in some other convenient manner, at the receiving end. From the received record, by suitable processes, already known and applied, hundreds of thousands of copies of the newspaper could be run off the presses. In this way, an exact copy of, say, this morning's London *Times* could be made available the same day to the inhabitants of Cape Town, Calcutta, or Melbourne. The development of such a scheme would inevitably alter considerably the contents and make-up of the newspaper concerned, so that it contained items of more direct interest to those communities which it was designed to serve. In this way the truly international newspaper may yet come into being, and the world will then become just that much smaller, and its scattered inhabitants knit just that much more closely together by this application of the latest and fastest means of communication known to man.

THE END.

INDEX.

A LEXANDERSON, E. F. W., 140, 189, 201-203, 226, 227, 233.
Alkali metals, 23, 80.
American Telephone and Telegraph Co., 87, 191.
Antimonite, 19.
Aperture distortion, 128-131.
Ardenne, Manfred von, 157, 168-169.
Argon, 26, 103, 157-158.
Aylesworth, M. H., 228.

B AIRD Co., 157, 170.
— John L., 58, 170.
— " Televisor," 175-179.
Beam scanning, 98-100.
Belin, 42-45.
Bell, Alexander Graham, 2, 14, 58.
— *System Tech. Journal*, 95, 196.
— Telephone Laboratories, 87, 90, 98, 105, 189, 191-201, 225-227, 230.
Berengaria, 139.
Berlin Radio Exhibition, 158, 164-165, 169, 193.
Beryllium, 194.
Bias voltage (neon), 96-98, 104-105.
Bosch, 157.
Braun tube, 36-38, 44-45, 97.
B.B.C., 151, 170-171.
Brookmans Park, 144, 170, 174-175.

C ÆSIUM, 23, 76, 80, 171.
Cathode glow, 49.
— rays, 74-75.
— ray tubes, 36-38, 44-45, 168, 169, 208-210, 213, 214-221.
Cells, light-sensitive, 13, 18-32.
Cellular scanner, 64.
Clarkson, R. P., 229.
Codel, Martin, 230.
Coliseum, 181.
Colour television, 79-82, 100-106.
Columbia Broadcasting System, 228.
Commutators, 93-95, 109-110, 180-182.
Control room (Baird's), 171-175.

Cox selenium relay, 28.
Crookes dark space, 49.

D 'ALBE, E. E. Fournier, 20
Daylight television, 78, 98-100.
DeForest, Lee, 230.
Direct scanning, 98-100.
Direction finding, 236.
Dissector tube, 214-218.
Distributor, 93-95.
Dot theory, 123-128.
Double images, 163-164, 186-187.
— modulation, 151-155.
Drum scanner, 54-57.

E CHO images, 163-164, 186-187.
Electron gun, 208-209, 220.
— image, 215.
Elster, 25.
Equalisers, 130.
Eye, electric, 58.
— human, 12-13.

F ARNSWORTH, PHILO T., 189, 213-226, 228.
Federal Radio Commission, 204.
Fernseh, A.G., 157, 169.
Filters, 76, 79, 100-104.
Fleming, Sir Ambrose, 143, 145.
Fluorescent screen, 36, 37, 74.
Focal point, 9, 10.
Focus, 10, 11.
Fournier, 15.
Fournier's theorem, 145.
Frith Street, 59.

G .E.C. (England), 23-26, 171.
— (U.S.A.), 189, 201-203, 227, 233.
Geitel, 25.
Geltow, 163.
German P.O., 157-159.
Ghost images, 163-164, 186-187.
Glow discharge, 48-49, 93-98.

Goldsmith, Alfred N., 229.
Green, Eldridge, 13.
Grid neon, 95-98.
Ground wave, 186-187, 223.

HALF-TONE process, 124-125.
Hallwachs, 23.
Hammond, John Hays, Jr., 151-155, 235.
Heaviside layer, 186.
Hefner, 168.
Heising, 97.
Helium, 50, 81.
Hertz, 23.
Holweck, 42-45.
Horn, C. W., 230.
Hunting, 51, 116, 119, 120.
Hydrogen, 194-195.

"IMAGE COMPRESSOR," 224.
— dissector, 214-218.
— exploring, 17, 35, 36-37, 39-45, 53-57, 60-70, 77-86, 98-100, 126-127, 132-134.
Images, virtual and real, 9, 11.
Incidence, angle of, 8, 159.
Infra-red rays, 74-78, 135.
Institute of Radio Engineers, 206.
International Radio Convention of 1927, 203.
Ionisation, 27, 48-49.
Isochronism, 115-116, 178.

JENKINS, C. FRANCIS, 46-57.
— Television Corp., 189-191, 226.
Jersey City, 189.
Jewett, Frank B., 230.

KAROLUS, 157, 159-164, 169, 201-203, 226.
Kerr effect, 161-162.
Kinescope, 209-211.
Koenigswusterhausen, 158.

LAMBERT'S cosine law, 8.
Large screen (Alexanderson), 201-203.
— (Baird), 181-183.
— (Bell), 93-98.
— (Karolus), 161-163.
Lens discs, 51, 53, 63.
Lenses, 10, 99-100.
Lightning, explanation of, 48.
Light relay, 40-42.
Lithium, 80.

Loewe Radio, 157.
Lumen, 23.
Lux, 20.

MAGNETIC focusing, 216-218.
May, 14.
Mercury vapour, 50, 81, 157-158.
Mihaly, Denes von, 38-42, 157, 164-169.
Mirrors, 8-9, 33-45, 80, 101-103.
Moore, D. MacFarlan, 50-51.
Multiplex transmission, 151-155.

NATIONAL Broadcasting Company, 227-228, 230.
Nature, 143, 145.
Nauen, 163.
Negative glow, 49, 55, 62.
— image, 174-175, 185-187.
Neon lamps, 47-51, 55, 62, 93-98, 110, 161, 194-195.
Nicol prisms, 161-163, 201-203.
Nipkow, 59-60, 63.
Nitrol-benzol, 162.
Noctovision, 73-78, 235.

OPTICAL lever, 64-69.
Optic nerve, 13.
Oscillite, 219-221.
Oscillograph, 39-42, 44-45.

PERISCOPE, 165.
Persistence of vision, 13-14, 17, 61, 82.
Phase displacement, 51, 113-115, 117, 120, 199.
— reversal, 149-151, 185-186.
Phasing, 116-120, 199.
Phonic wheel, 39-41, 111-112, 164-167, 169, 176-179.
Phonovision, 71-73.
Photo-conductivity, 18-19.
Photo-electric cells, 13, 20, 23-30, 49, 80, 92, 100-102, 193.
— — disposition of, 92-93, 101-102.
Photo-emissivity, 18, 23, 28.
Phototelegraphy, 4-5, 90, 107-110, 113-115, 121-123, 236-237.
Picture elements, 125-128, 160, 193, 232.
— ratio, 131-133.
Piezo-electric crystals, 110-111.
Polarisation, 161-162.
Positive column, 49-51.
Potassium, 23, 26, 30, 80, 100-101, 193.

INDEX

Prismatic disc, 46-47.
Prisms, 11, 12, 161-163.

RADIO City, 228.
— Corporation of America, 108, 201, 227-229.
— Manufacturers' Association (U.S.A.), 205, 229.
Radio-Keith-Orpheum, 227.
Radiomovie, 55.
Radiovisor Bridge, 20-23.
Ranger, R. H., 108-110.
R.C.A. Victor Co., 189, 206, 227.
Record transmissions, 138, 139-140.
Reflection, angle of, 8, 159.
Refraction, 7, 10.
Relative motion, 67.
Retentivity of vision, 13.
Retina, 13, 15.
Rignoux, 15.
Robinson, J., 125, 139, 144-151, 236.
Rodwin, 111.
Rosing, Boris, 36-38.
Royal Institution, 59.
Rubidium, 23, 80.
Ruhmer, Ernest, 16.

SANABRIA, U. A., 69.
Savoy Hill, 173-175.
Scanning, 32, 35, 36-37, 39-47, 53-57, 60-70, 77-86, 98-100, 126-127, 132-134, 207.
Schenectady, 201, 233.
Science Museum, South Kensington, 59.
Selenium, 14-16, 18-23, 28.
Selfridge's, 59.
Shadowgraphs, 31, 48.
Shutters, 15-16.
Sideband theory, 140-146.
Siemens oscillograph, 39-40.
Signal bias (neon), 96-98.
— frequency, 135-137, 177.
Signalling speeds, 121-123.
Smith, 111.
— Willoughby, 18.
Sodium, 23, 80, 100-102.
Space wave, 186-187.
Spectrum, 74-75.
Sperry arc, 90.
Steatite, 18.
Stenode Radiostat, 144-151, 236.
Stereoscopic television, 82-84.

Stroboscopic effect, 110, 117.
Studio (Baird's), 171.
Sulphur, 19.
Synchronism, 6, 36, 45, 51, 56, 107-120, 160-161, 165-168, 176-181, 185, 190, 195-199, 210-211, 221-222.
Synchronous motors, 51, 56, 111-113, 116-120, 160-161, 203.
Szczepanik, Jan van, 33-36.

TELEFUNKEN, 157, 159-164, 169.
Tele-talkies, 181, 183, 184.
Television (definition), 1-2.
— by daylight, 78, 98-100.
— in colours, 79-82, 100-106.
— Laboratories, Inc., 189, 213.
— (requirements), 5-6, 58:
— stereoscopic, 82-84.
Thallium oxysulphide, 19.
Time lag, 19, 23, 28.
Tuning-forks, 39-41, 51, 108-110, 160-161.
Two-way television, 158, 191-201.

ULTRA-VIOLET rays, 74-76.

VISION, persistence of, 13-14, 17, 61, 82.
Visual purple, 13, 15.

WEILLER mirror wheel, 159-160.
Weinberger, 111.
Westinghouse Electric and Mfg. Co., 30, 189, 206, 227.
WGBS, 189.
Willemite, 208.
Wired wireless, 223.
Witzleben, 158.

X-RAYS, 49, 74-75.

YOUNG, Owen D., 233.

ZEISS-Ikon, 157.
Zworykin, V. K., 30, 189, 206-213, 226, 227.

HISTORY OF BROADCASTING:
Radio To Television
An Arno Press/New York Times Collection

Archer, Gleason L.
Big Business and Radio. 1939.

Archer, Gleason L.
History of Radio to 1926. 1938.

Arnheim, Rudolf.
Radio. 1936.

Blacklisting: Two Key Documents. 1952–1956.

Cantril, Hadley and Gordon W. Allport.
The Psychology of Radio. 1935.

Codel, Martin, editor.
Radio and Its Future. 1930.

Cooper, Isabella M.
Bibliography on Educational Broadcasting. 1942.

Dinsdale, Alfred.
First Principles of Television. 1932.

Dunlap, Orrin E., Jr.
Marconi: The Man and His Wireless. 1938.

Dunlap, Orrin E., Jr.
The Outlook for Television. 1932.

Fahie, J. J.
A History of Wireless Telegraphy. 1901.

Federal Communications Commission.
Annual Reports of the Federal Communications Commission. 1934/1935–1955.

Federal Radio Commission.
Annual Reports of the Federal Radio Commission. 1927–1933.

Frost, S. E., Jr.
Education's Own Stations. 1937.

Grandin, Thomas.
The Political Use of the Radio. 1939.

Harlow, Alvin.
Old Wires and New Waves. 1936.

Hettinger, Herman S.
A Decade of Radio Advertising. 1933.

Huth, Arno.
Radio Today: The Present State of Broadcasting. 1942.

Jome, Hiram L.
Economics of the Radio Industry. 1925.

Lazarsfeld, Paul F.
Radio and the Printed Page. 1940.

Lumley, Frederick H.
Measurement in Radio. 1934.

Maclaurin, W. Rupert.
Invention and Innovation in the Radio Industry. 1949.

Radio: Selected A.A.P.S.S. Surveys. 1929–1941.

Rose, Cornelia B., Jr.
National Policy for Radio Broadcasting. 1940.

Rothafel, Samuel L. and Raymond Francis Yates.
Broadcasting: Its New Day. 1925.

Schubert, Paul.
The Electric Word: The Rise of Radio. 1928.

Studies in the Control of Radio: Nos. 1–6. 1940–1948.

Summers, Harrison B., editor.
Radio Censorship. 1939.

Summers, Harrison B., editor.
A Thirty-Year History of Programs Carried on National Radio Networks in the United States, 1926–1956. 1958.

Waldrop, Frank C. and Joseph Borkin.
Television: A Struggle for Power. 1938.

White, Llewellyn.
The American Radio. 1947.

World Broadcast Advertising: Four Reports. 1930–1932.